Johannes Schranck

Transition Metal-Catalyzed Carbonylative Coupling Reactions

AF061128

Johannes Schranck

Transition Metal-Catalyzed Carbonylative Coupling Reactions

Synthesis of Aromatic Carbonyl Compounds

Südwestdeutscher Verlag für Hochschulschriften

Impressum / Imprint
Bibliografische Information der Deutschen Nationalbibliothek: Die Deutsche Nationalbibliothek verzeichnet diese Publikation in der Deutschen Nationalbibliografie; detaillierte bibliografische Daten sind im Internet über http://dnb.d-nb.de abrufbar.
Alle in diesem Buch genannten Marken und Produktnamen unterliegen warenzeichen-, marken- oder patentrechtlichem Schutz bzw. sind Warenzeichen oder eingetragene Warenzeichen der jeweiligen Inhaber. Die Wiedergabe von Marken, Produktnamen, Gebrauchsnamen, Handelsnamen, Warenbezeichnungen u.s.w. in diesem Werk berechtigt auch ohne besondere Kennzeichnung nicht zu der Annahme, dass solche Namen im Sinne der Warenzeichen- und Markenschutzgesetzgebung als frei zu betrachten wären und daher von jedermann benutzt werden dürften.

Bibliographic information published by the Deutsche Nationalbibliothek: The Deutsche Nationalbibliothek lists this publication in the Deutsche Nationalbibliografie; detailed bibliographic data are available in the Internet at http://dnb.d-nb.de.
Any brand names and product names mentioned in this book are subject to trademark, brand or patent protection and are trademarks or registered trademarks of their respective holders. The use of brand names, product names, common names, trade names, product descriptions etc. even without a particular marking in this work is in no way to be construed to mean that such names may be regarded as unrestricted in respect of trademark and brand protection legislation and could thus be used by anyone.

Coverbild / Cover image: www.ingimage.com

Verlag / Publisher:
Südwestdeutscher Verlag für Hochschulschriften
ist ein Imprint der / is a trademark of
OmniScriptum GmbH & Co. KG
Heinrich-Böcking-Str. 6-8, 66121 Saarbrücken, Deutschland / Germany
Email: info@svh-verlag.de

Herstellung: siehe letzte Seite /
Printed at: see last page
ISBN: 978-3-8381-3951-7

Zugl. / Approved by: Rostock, Universität Rostock, Diss., 2014

Copyright © 2014 OmniScriptum GmbH & Co. KG
Alle Rechte vorbehalten. / All rights reserved. Saarbrücken 2014

Acknowledgement

Mein besonderer Dank gilt meinem Doktorvater *Prof. Dr. Matthias Beller* für die Bereitstellung des interessanten Themas, die ausgezeichneten Arbeitsbedingungen, sowie die stetige freundliche Unterstützung. Weiterhin bin ich ihm für die unerschöpflichen wertvollen Anregungen, viele gewährte Freiheiten und Möglichkeiten, sein Vertrauen und sein großes Interesse am Gelingen dieser Arbeit sehr verbunden.

Außerdem danke ich meinem Themenleiter *Dr. Helfried Neumann* für die nette Betreuung und permanente Hilfsbereitschaft bei der Lösung von Fragen und Problemen rund um das bearbeitete Thema und darüber hinaus.

I am deeply grateful to *Dr. Anis Tlili* for his generous and extensive support, the numerous successful collaborations, fruitful discussions and a cordial friendship.

Ich danke der gesamten Themengruppe „Übergangsmetall-katalysierte Synthesen von Feinchemikalien" für sämtliche fachliche und persönliche Hinweise, Ratschläge und die nette Arbeitsatmosphäre. Insbesondere möchte ich mich bei *Dr. Andreas Dumrath* für die geduldige Unterstützung und die vielen freundschaftlichen Konsultationen bedanken. Für das ausgesprochen gute Arbeitsklima und die stete Hilfsbereitschaft danke ich *Dr. Sebastian Bähn, Dr. Saravanan Gowrisankar, Dr. Sebastian Imm, Christa Lübbe, Dr. Kishore Natte, Dr. Miguel Peñja-Lopez, Jola Pospech* und *Dr. Xiao-Feng Wu*.

Acknowledgement

Sandra Leiminger danke ich ganz besonders für die sehr angenehme Zusammenarbeit und vielfältige Unterstützung im Laboralltag und darüber hinaus.

I am greatly indebted to *Prof. Dr. Mark Stradiotto* for hosting me in his research group at Dalhousie University, Halifax, Canada and for giving me the opportunity to work in his research environment. I am deeply grateful for the strong support and the very convenient collaboration I experienced in his research group. Particularly, I would also like to thank *Dr. Pamela G. Alsabeh*, *Dr. Christopher Lavery*, *Dr. Craig Wheaton* and *Dr. Sarah Crawford* for their very warm welcome, their support during my stay, many scientific and intercultural conversations and a great summer in Canada.

I also thank *Mia Burhardt* and *Prof. Dr. Troels Skrydstrup* for a very fruitful collaboration and their support.

Der gesamten Arbeitsgruppe danke ich für die gute Zusammenarbeit und die freundliche Gemeinschaftlichkeit bei Seminaren, Mensabesuchen und sonstigen Ereignissen.

Für die schöne Zeit am Leibniz-Institut für Katalyse e.V. danke ich allen Freunden und Kollegen, insbesondere *Jenny Bandomir*, *Christoph Bornschein*, *Hendrik Büttner*, *Christoph Cordes*, *Johannes Diebler*, *Eric Enderle*, *Dr. Christopher Federsel*, *Dr. Steffen Fleischer*, *Dr. Felix Gärtner*, *Dr. Marko Hapke*, *Dr. Ralf Jackstell*, *Dr. Haijun Jiao*, *Dr. Benoit Join*, *Dr. Henrik Junge*, *Dr. Kathrin Junge*, *Dr. Katharina Kaleta*, *Dr. Yuehui Li*, *Dr. Henrik Lund*, *Dr. Angèle Monney*, *Dr. Anahit Pews-Davtyan*, *Dr. Nils Rockstroh*, *Dr. Thomas Schareina*, *Peter Sponholz*, *Bianca Wendt*, *Lipeng Wu* und vielen mehr.

Acknowledgement

Der analytischen Abteilung um *Dr. Wolfgang Baumann*, *Susanne Buchholz*, *Dr. Christine Fischer*, *Andreas Koch*, *Dr. Dirk Michalik* und *Susanne Schareina* danke ich für die zahlreichen durchgeführten Messungen, sowie *Grit Apportin*, *Dr. Torsten Dwars*, *Torsten Weiss* und *Claudia Zielke* für die schnelle Chemikalienbeschaffung und die unkomplizierte Zusammenarbeit.

Den Kollegen aus der Verwaltung danke ich ebenfalls für die stets freundliche und reibungslose Zusammenarbeit, insbesondere *Kerstin Gasch*, *Anke Kirmse*, *Heike Koch*, *Stefan Legler*, *Jenny Rehbein* und *Dr. Johannes Treu*.

Ganz besonderer Dank gilt meiner Familie, insbesondere meinen Eltern *Kathrin* und *Torsten*, die mir durch ihre überragende Integrität und nicht zuletzt durch ihre umfangreiche Unterstützung die Möglichkeit gaben, mein Studium erfolgreich zu absolvieren und diese Arbeit zu verfassen.

I am particularly grateful for the support and encouragement given by *Frances Sweeney*.

4

Abstract

Transition Metal-Catalyzed Carbonylative Coupling Reactions
Johannes Schranck
Leibniz-Institut für Katalyse e.V. an der Universität Rostock

This thesis describes the development of novel transition metal-catalyzed carbonylative C–C bond-forming reactions with a focus on the employment of (activated) arenes in combination with simple unfunctionalized carbon nucleophiles in order to prevent stoichiometric amounts of metal waste. This concept is realized in two examples of new palladium-catalyzed carbonylative Heck-type reactions between aryl halides and vinyl ethers or terminal olefins producing 3-alkoxyalkenones or 5-arylfuranones, respectively. Furthermore, three protocols have been developed for the palladium-catalyzed carbonylative coupling of aryl iodides with C–H acidic compounds, that is the coupling of benzyl ketones to vinyl benzoates, the coupling of acetone and acetophenones to 1,3-diketones and the coupling of nitriles to β-ketonitriles. In a third approach, the ruthenium-catalyzed C–H bond functionalization of arenes has been utilized to conduct carbonylative coupling with aryl iodides yielding benzophenones and has allowed for the development of the hydroaroylation of styrenes to form the corresponding saturated ketones.

Die vorliegende Arbeit beschreibt die Entwicklung neuer Übergangsmetall-katalysierter carbonylierender C–C-Bindungsknüpfungen mit Fokus auf dem Einsatz von (aktivierten) Aromaten in Kombination mit nicht vorab funktionalisierten Kohlenstoffnukleophilen, um stöchiometrische Mengen metallhaltiger Abfälle zu vermeiden. In zwei Beispielen wird dieses Konzept in neuen Varianten der

Palladium-katalysierten carbonylierenden Heck-Reaktion von Arylhalogeniden mit Vinylethern oder terminalen Olefinen realisiert, wobei 3-Alkoxyalkenone bzw. 5-Arylfuranone gebildet werden. Zudem wurden drei weitere Systeme für die Palladium-katalysierte carbonylierende Kupplung von C–H-aciden Verbindungen entwickelt. Dabei reagieren Benzylketone zu Vinylbenzoaten, Aceton und Acetophenone zu 1,3-Diketonen sowie Nitrile zu den entsprechenden β-Ketonitrilen. Ein dritter Ansatz nutzt die Ruthenium-katalysierte C–H-Funktionalisierung von Aromaten, um die carbonylierende Kupplung mit Aryliodiden zu Benzophenonen sowie die Hydroaroylierung von Styrolen zu den entsprechenden gesättigten Ketonen durchzuführen.

Table of Contents

Acknowledgement 1

Abstract 5

Table of Contents 7

List of Schemes 11

List of Abbreviations 15

1　Preface 19

2　Carbonylative Coupling Reactions 23

　2.1 Three Component Carbonylative Heck Reaction 27

　2.2 Carbonylative α-Arylations 36

　2.3 Carbonylative C–C Bond Formation via C–H Bond Activation of Arenes 46

　　2.3.1　Carbonylative C–C Cross Coupling via C–H Bond Activation of Arenes 47

　　2.3.2　Carbonylative Hydroarylation via Directing Group Assisted C–H Bond Activation of Arenes 51

3　Objectives of This Work 58

4　Summary of Results 60

　4.1 Carbonylative Heck Reaction of Aryl Bromides with Vinyl Ethers 60

　4.2 Doublecarbonylation Reaction of Aryl Halides to Arylfuranones 65

　4.3 Carbonylative Arylation of Aryl Ketones to Vinylbenzoates 70

　4.4 Carbonylative α-Arylation of Acetone and Acetophenones 73

　4.5 Carbonylative α-Arylation of Nitriles 78

4.6 Carbonylative Transformations of Bromhexine 83

4.7 Aroylation of Aryl Iodides via Directed C–H Bond Activation 86

4.8 Hydroaroylation of Styrenes via Directed C–H Bond Activation 90

4.9 Iron-Catalyzed Selective Reduction of Aldehydes under Water-Gas Shift Conditions 95

4.10 Conclusion 100

5. Publications 102

5.1 Palladium-Catalyzed Carbonylative Heck Reaction of Aryl Bromides with Vinyl Ethers to 3-Alkoxy Alkenones and Pyrazoles 102

5.2 A Novel Double Carbonylation Reaction of Aryl Halides: Selective Synthesis of 5-Arylfuranones 103

5.3 A Selective Palladium-Catalyzed Carbonylative Arylation of Aryl Ketones to Give Vinylbenzoate Compounds 104

5.4 Palladium-Catalyzed Carbonylative α-Arylation of Acetone and Acetophenones to 1,3-Diketones 105

5.5 Palladium-Catalyzed Carbonylative α-Arylation to β-Ketonitriles 106

5.6 Palladium-Catalyzed Carbonylative Transformations of Bromhexine into Bioactive Compounds as Glucocerebrosidase Inhibitors 107

5.7 Ruthenium-Catalyzed Carbonylative C–C Coupling in Water by Directed C–H Bond Activation 108

5.8 Ruthenium-Catalyzed Hydroaroylation of Styrenes in Water via Directed C–H Bond Activation 109

5.9 Discrete Iron Complexes for the Selective Catalytic Reduction of Aromatic, Aliphatic, and α,β-Unsaturated Aldehydes under Water-Gas Shift Conditions 110

Table of Contents

6. References 111

10

List of Schemes

Scheme 1: Chemical production, raw material use, energy consumption, emissions. 19
Scheme 2: Carbonylation of aryl (pseudo)halides. 23
Scheme 3: Palladium-catalyzed carbonylation of aryl halides – catalytic cycle. 24
Scheme 4: Products of classical carbonylative cross-coupling reactions. 25
Scheme 5: Carbonylative cross-coupling of aryl iodides with 2,3-dihydrofuranes. 28
Scheme 6: Carbonylative cross-coupling of 2-iodophenoles with allenes. 28
Scheme 7: Carbonylative cross-coupling of 2-iodoanilines with allenes. 29
Scheme 8: Addition of in situ generated molybdenum-acyl complexes onto alkenes. 29
Scheme 9: Carbonylative cross-coupling of aryl/vinyl triflates with styrenes. 30
Scheme 10: Carbonylative cross-coupling of aryl halides with olefins. 31
Scheme 11: Carbonylative cross-coupling of aryl bromides with styrenes. 32
Scheme 12: Catalytic cycle for the carbonylative Heck reaction of phenyl halides with styrene .. 33
Scheme 13: Carbonylative Heck reaction with styrenes using near-stoichiometric amounts of CO. 34
Scheme 14: Carbonylative Heck reaction with vinyl ethers using near-stoichiometric amounts of CO. 35
Scheme 15: Carbonylative coupling of aryl iodides with carbonyl derivatives. 38
Scheme 16: Intramolecular carbonylative α-arylation/vinylation of activated esters. 39
Scheme 17: Intramolecular carbonylative α-arylation of ketones. 40
Scheme 18: Carbonylative coupling of aryl iodides with ethyl diazoacetate. 41
Scheme 19: Intermolecular carbonylative α-arylation of ketones. 42
Scheme 20: Carbonylative α-arylation of 1,3-diketones to triketones. 42
Scheme 21: Carbonylative α-arylation of 1,3-diketones with subsequent deacetylation. 43
Scheme 22: Carbonylative α-arylation of monoester potassium malonates. 44
Scheme 23: Intramolecular cyclocarbonylation via C–H activation of arenes. 48
Scheme 24: Carbonylative cross-coupling of aryl iodides with heteroarenes. 49
Scheme 25: Intramolecular oxidative double C–H bond functionalization of diarylethers. 50
Scheme 26: Reaction of pyridine with carbon monoxide and 1-hexene. 51
Scheme 27: Reaction of pyridylbenzenes with carbon monoxide and ethylene. 53
Scheme 28: Proposed catalytic cycle for the reaction of 2-pyridylbenzene with CO and ethylene. 54
Scheme 29: Reaction of aza-heterocycles with carbon monoxide and alkenes. 55

List of Schemes

Scheme 30: Oxazoline and pyridine acting as directing groups. 56
Scheme 31: Reaction of N-pyridylindolines with carbon monoxide and alkenes. 57
Scheme 32: Variation of ligands in the carbonylative Heck reaction of vinyl ethers. 61
Scheme 33: Variation of aryl bromides in the carbonylative Heck reaction with vinyl ethers 62
Scheme 34: Variation of vinyl ethers in the carbonylative Heck reaction with aryl bromides. 63
Scheme 35: One-pot synthesis of pyrazoles 63
Scheme 36: Carbonylative Heck coupling of PhOTf, PhI and PhBr with 1-octene. 66
Scheme 37: Palladium-catalyzed doublecarbonylative coupling of aryl iodides and bromides with 1-octene. 67
Scheme 38: Palladium-catalyzed doublecarbonylative coupling of bromobenzene with terminal alkenes. 68
Scheme 39: Proposed mechanism for the doublecarbonylative coupling of bromobenzene with 1-octene. 69
Scheme 40: Palladium-catalyzed carbonylative coupling of aryl bromides and iodides with deoxybenzoin. 71
Scheme 41: Palladium-catalyzed carbonylative coupling of iodobenzene with different aryl ketones. 72
Scheme 42: Palladium-catalyzed carbonylative coupling of acetone with different aryl iodides... 75
Scheme 43: Palladium-catalyzed carbonylative α-arylation of different ketones to 1,3-diketones. 76
Scheme 44: One-pot synthesis of pyrazoles via carbonylative α-arylation of ketones. 77
Scheme 45: Palladium-catalyzed carbonylative coupling of nitriles with different aryl iodides. 80
Scheme 46: Palladium-catalyzed carbonylative coupling of 3-iodotoluene with different nitriles.. 81
Scheme 47: Carbonylative α-arylation of isobutyronitrile with ^{13}COgen. 82
Scheme 48: Selective reduction of β-ketonitrile to β-hydroxynitrile. 82
Scheme 49: Palladium-catalyzed carbonylative coupling of bromhexine with arylboronic acids. .. 84
Scheme 50: Palladium-catalyzed carbonylative coupling of bromhexine with alcohols and amines. 84
Scheme 51: Inhibition of recombinant GCase by bromhexine and chosen compounds 85
Scheme 52: Ruthenium-catalyzed carbonylative C–H functionalization with different aryl iodides. 87
Scheme 53: Ruthenium-catalyzed carbonylative C–H functionalization with different directing groups. 88
Scheme 54: Ruthenium-catalyzed aroylation of aryl iodides – deuteration experiments. 89
Scheme 55: Ruthenium-catalyzed hydroaroylation of different styrene derivtatives 91
Scheme 56: Ruthenium-catalyzed hydroaroylation of alkenes. 92

List of Schemes

Scheme 57: Ruthenium-catalyzed hydroaroylation of pentafluorostyrene with different directing groups. *93*

Scheme 58: Ruthenium-catalyzed hydroaroylation of pentafluorostyrene – deuteration experiments. *94*

Scheme 59: Reduction of benzaldehyde to benzyl alcohol – testing of different iron carbonyl complexes. *97*

Scheme 60: Iron-catalyzed reduction of aromatic aldehydes to alcohols via WGSR. *98*

Scheme 61: Iron-catalyzed reduction of aliphatic aldehydes to alcohols via WGSR. *98*

14

List of Abbreviations

1-Ad	1-adamantyl
Ac	acetyl
Ar	aryl
BINAP	2,2'-bis(diphenylphosphino)-1,1'-binaphthyl
bmim	1-*n*-butyl-3-methylimidazolium
Bn	benzyl
cata*CX*ium A	di-1-adamantyl-*n*-butylphosphane
cata*CX*ium POMeCy	1-(2-methoxyphenyl)-2-(dicyclohexylphosphino)pyrrole
cod	cycloocta-1,5-diene
Cy	cyclohexyl
dba	*trans,trans*-dibenzylideneacetone
DFT	density functional theory
DiPEA	*N,N*-diisopropylethylamine
DiPPF	1,1'-bis(diisopropylphosphino)ferrocene
DMAc	dimethylacetamide
DMF	*N,N*-dimethylformamide
dppb	1,4-bis(diphenylphosphino)butane
dppf	1,1'-bis(diphenylphosphino)ferrocene
dppp	1,3-bis(diphenylphosphino)propane
dppy	2-(diphenylphosphino)pyridine
equiv.	equivalent(s)
Et	ethyl
GC	gas chromatography
HBF_4	fluoroboric acid

i-Pr	isopropyl
KO*t*Bu	potassium-*tert*-butoxide
HMDS	hexamethyldisilazane
L	ligand
LDA	lithium diisopropylamide
m-, *o*-, *p*-	*meta*-, *ortho*-, *para*-
Me	methyl
Mes	mesityl (2,4,6-trimethylphenyl)
NaO*t*Bu	sodium-*tert*-butoxide
Nf	nonaflate
NMP	*N*-methyl pyrrolidone
NMR	nuclear magnetic resonance
Np	naphthyl
Nu	nucleophile
OAc	acetate
OMe	methoxy
OMs	mesylate (methanesulfonate)
OTf	triflate (trifluoromethanesulfonate)
o-tol	*ortho*-tolyl
OTs	tosylate (*p*-toluenesulfonate)
Ph	phenyl
Pin	pinacol
PPh$_3$	triphenylphosphane
p-TsOH	*p*-toluenesulfonic acid
Py	pyridine
rf	reflux
rt	room temperature
SPhos	2-dicyclohexylphosphino-2',6'-dimethoxybiphenyl

List of Abbreviations

TBAF	tetra-*n*-butylammonium fluoride
TMS	trimethylsilyl
TFA	trifluoroacetate
THF	tetrahydrofuran
TMEDA	*N,N,N',N'*-tetramethylethylenediamine
TOF	turnover frequency; [TOF] = 1 h^{-1}
TON	turnover number
WGSR	water-gas shift reaction

18

1 Preface

In 2010, the 434,000 employees of Germany's chemical industry accounted for more than 186 billion euro turnover.[1] Being among the five most powerful branches of industry, the German chemical industry has a significant impact on the affluence of our country. Due to the lack of natural resources, Germany has a constant interest in developing highly efficient technologies for generating value from imported raw materials. Energy consumption and greenhouse gas emission normally increase with volume of chemical production. However, the German Chemical Industry Association (VCI - Verband der Chemischen Industrie) has reported a successful inverse to this trend between 1990 and 2010, which demonstrates that energy consumption and greenhouse gas emission can be reduced with increased chemical production.[2]

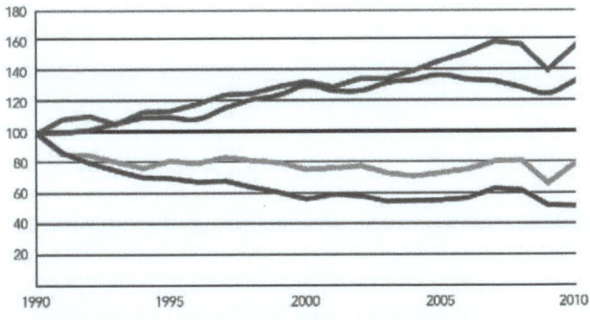

- Production index of the chemical/pharmaceutical industry
- Petrochemical raw materials index (use of petrochemical raw materials)
- Energy consumption index (total amount of coal, heating oil, natural gas and net electricity purchased)
- Greenhouse gas emissions index (energy-related CO_2 and NO_2 emissions)

Scheme 1: Chemical production, raw material use, energy consumption, emissions.

Thus, in no other region on Earth are chemicals manufactured as energy efficient and climate-friendly as in the Federal Republic of Germany. To highlight chemistry's impact on the world and to gain public focus on future challenges of sustainable development, UNESCO deemed 2011 the "International Year of Chemistry."[3] In this context, catalysis plays a key role in gaining the economic as well as ecological advantages of sustainable chemistry. As the science of accelerating chemical processes, catalysis is of significant importance to industry as well as academia. Depending on the line of business, each euro invested into catalysis can generate added values between 100 and 400 €.[4] Consequently, catalysts are currently involved in more than 80% of all manufacturing processes in the chemical industry.[5] This results in a persistent interest in the development of new catalysts and in applying them to novel synthetic methodologies.[6]

Especially in the area of fine chemicals, homogeneous catalysis has become an essential and powerful tool for synthetic organic chemists. Offering high activities and superior selectivities under mild reaction conditions, homogeneous catalysis has outperformed many established stoichiometric reaction systems.[7] Thus, modern homogeneous catalysts find a plethora of applications i.e. hydroformylation, hydrogenation, oxidation, metathesis, as well as carbonylation and cross-coupling reactions.[8] The latter have become a significant tool for the formation of C–C, C–N, C–O, or C–S bonds, being excessively applied in both academic and industrial settings.[9]

Pioneering work in this area was conducted by Mizoroki and Heck in the early 1970s when they independently achieved the first catalytic cross coupling reactions by reacting aryl halides with olefins in the presence of a palladium catalyst.[10] Since then, C–C cross-coupling has been an intensively studied field in homogeneous catalysis and has led to an entire toolbox of

1 Preface

methodologies, with the application of different carbon nucleophiles being key to versatility. In addition to olefins that were used by Mizoroki and Heck,[11] Sonogashira and Hagihara discovered alkynes as suitable coupling partners.[12] In the following decades, further reagents such as (hetero)aryl, vinyl or alkyl derivatives of boron (Suzuki, Miyaura),[13] zinc (Negishi),[14] tin (Stille, Migata),[15] magnesium (Kumada, Corriu),[16] zirconium (Negishi)[17] and silicon (Hiyama)[18] were found to be suitable coupling partners for coupling with a variety of (hetero)aryl and alkyl (pseudo)halides. More recently, the groups of Miura,[19] Buchwald[20] and Hartwig[21] concurrently reported intermolecular α-arylation reactions of ketones. As a result, such α-arylations have been extended to the use of a variety of C–H acidic moieties such as esters,[22] aldehydes,[23] amides[24] or nitriles[25] as nucleophiles in C–C coupling reactions.

By introducing one molecule of carbon monoxide into a target molecule, carbonylation reactions have become a very useful reaction in bulk as well as in fine chemical applications. Thus, carbon monoxide is amongst the most important C1 building blocks in the chemical industry, especially for hydroformylation reactions that allow for the large-scale production of alcohols, aldehydes and carboxylic acid derivatives.[26] With respect to fine chemical synthesis, carbonylation reactions of aryl (pseudo)halides in particular have developed as a widespread field in organic synthesis, most often relying on the outstanding performance of homogeneous catalysts.[27]

As mentioned above, homogeneous catalysts, most often metal complexes that are dissolved in the same phase as the reactants, play a key role in coupling reactions and carbonylations, and thus will be predominant in the procedures discussed herein.[8] For general cross-coupling reactions, transition metal complexes of palladium, nickel and copper have been shown to be very effective,[9] while carbonylations are typically catalyzed by complexes based on rhodium, ruthenium, cobalt or palladium.[26] Due to its abilities to catalyze both processes (i.e. coupling reactions and carbonylations), palladium has become the predominant metal for conducting carbonylative coupling reactions.[28]

2 Carbonylative Coupling Reactions

In combining carbonylation and coupling reaction technologies, the first examples were once again demonstrated by Heck and co-workers in 1974 when they reported the palladium catalyzed alkoxy-, amino-, and hydroxycarbonylation of iodobenzene forming the corresponding esters, amides and acids, currently also referred to as Heck carbonylation reactions.[29] This was the starting point for a variety of carbonylative coupling reactions between aryl (pseudo)halides and nucleophiles (Scheme 2). Thus, the application of different nucleophiles opened the field towards the simple synthesis of (hetero)aromatic and vinylic aldehydes, esters, amides, ketones, or carboxylic acids from the same parent aryl (pseudo)halide.[30]

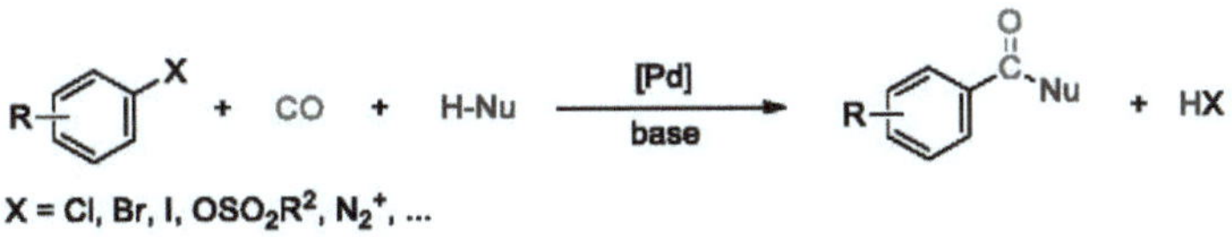

$$R\text{-}Ar\text{-}X + CO + H\text{-}Nu \xrightarrow[\text{base}]{[Pd]} R\text{-}Ar\text{-}C(O)\text{-}Nu + HX$$

$X = Cl, Br, I, OSO_2R^2, N_2^+, ...$

Scheme 2: Carbonylation of aryl (pseudo)halides.

Typically, these carbonylations are conducted in the presence of a palladium catalyst involving common organometallic elementary steps (Scheme 3).[31] The catalytic cycle starts with the oxidative addition of the aryl halide to the electronically unsaturated palladium(0) complex (**1**). Once the aryl palladium(II) complex is formed, the reversible insertion of CO into the aryl-palladium bond occurs (**2**). Subsequently, the nucleophile attacks the palladium-acyl complex (**3**) releasing HX, which is captured by stoichiometric amounts of base. Finally, reductive elimination enables product formation and regeneration of the catalytically active palladium(0) complex (**4**).

While the application of heteroatom-based nucleophiles has been intensively studied in the past, the use of carbon-based nucleophiles has been investigated less so. Hence, this thesis focuses on carbonylative C–C bond formations.

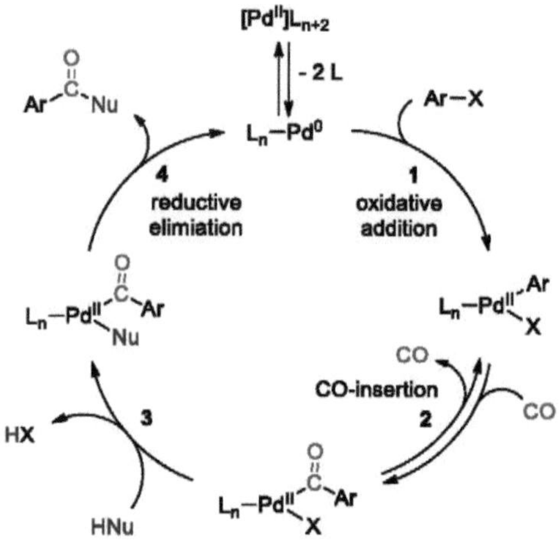

Scheme 3: Palladium-catalyzed carbonylation of aryl halides – catalytic cycle.

Since (hetero)aromatic and carbonyl compounds are vital intermediates in the manufacture of agrochemicals, dyes, pharmaceuticals, and other industrial products, their synthesis is of particular importance. In this context, carbonylative cross-coupling allows for the straightforward synthesis of complex carbonyl moieties from rather simple and easily available starting materials (Scheme 4). Conducting the reaction as a three-component reaction with gaseous carbon monoxide offers several advantages. Most importantly, the use of acid chlorides as starting material can be avoided;

these reagents often lack stability and also the commercial availability of their substituted derivatives is rather low, compared to (hetero)aryl (pseudo)halides. Furthermore, carbon monoxide constitutes a cheap and abundant C1 building block, albeit its associated toxicity requires appropriate equipment and caution.

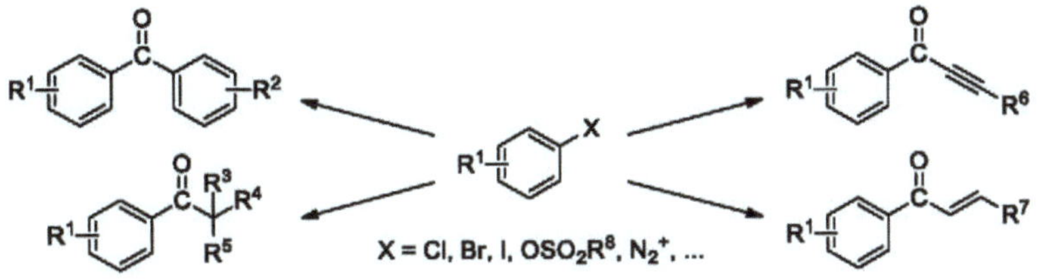

Scheme 4: Products of classical carbonylative cross-coupling reactions.

Inspiration for conducting carbonylative C–C cross-coupling reactions has come from their non-carbonylative cross-coupling correspondents. Hence, organotin reagents were the first to be used by Tanaka in 1979 for the synthesis of asymmetric ketones as an extension of the well-known Stille reaction.[32] Shortly after, also the first carbonylative Sonogashira coupling was reported, thus opening the gate for using alkynes as nucleophiles in order to assemble alkynylketones.[33] In the following decades, organoboron,[34] zinc,[35] and silicon[36] derivatives were also shown to be suitable coupling partners in carbonylation reactions of (hetero)aryl halides. After some preliminary intramolecular protocols had been reported,[37] Miura and co-workers published in 1995 the first example of an intermolecular carbonylative Heck reaction using 2,3-dihydrofuran as an alkene coupling partner.[38] In addition, there are some reports which have shown that carbonylative coupling can also be performed via C–H activation either intramolecularly on arenes[39] or intermolecularly on activated α-C–H acidic carbonyl compounds[40] as well as heteroarenes.[41]

Although the above-mentioned transformations offer interesting possibilities for organic chemistry, most of them require the use of (over)stoichiometric amounts of organometallic coupling reagents. The necessity to prepare these beforehand as well as the generation of stoichiometric amounts of metal waste lowers their synthetic value. Therefore, coupling carbon nucleophiles via functionalization of C–H bonds will be the main issue in the following sections.

2 Carbonylative Coupling Reactions

2.1 Three Component Carbonylative Heck Reaction

While carbonylative variants of several coupling reactions have already been developed since the early 1980s, only a few examples of analogous three-component Heck reactions had been reported at that time. In comparison to direct Heck coupling reactions, the presence of CO increases the demands imposed on the catalytic system in carbonylative coupling procedures. CO acts as a strong π-acceptor and a moderate σ-donor ligand, thus lowers the electron density of the metal center and in turn impedes the initial oxidative addition.[42] The application of suitable ligands is mandatory in order to secure an active catalytic system. Also, the formation of carbonyl palladium clusters is known to slow the rate of the overall reaction at high CO pressures. However, this can be suppressed by addition of extra phosphine ligand.

Although a variety of intramolecular carbonylative Heck-type reactions have been reported,[37] their intermolecular analogous have turned out to be more challenging and will be primarily discussed here. As mentioned above, Miura and co-workers were the first to realize an intermolecular carbonylative Heck reaction in 1995.[38] At 120°C and 5 bar of CO aryl iodides and dihydrofurans were reacted to give the corresponding ketones in yields ranging from 38 to 81%. (Scheme 5). Attempts to employ cyclopentene in their protocol generated the corresponding benzoylcyclopentene as a mixture of three regioisomers. The amount of triphenylphosphine ligand used was found to have a significant influence on product formation, that is, Pd:PPh$_3$ ratios higher than 1.0 resulted in decreased yields.

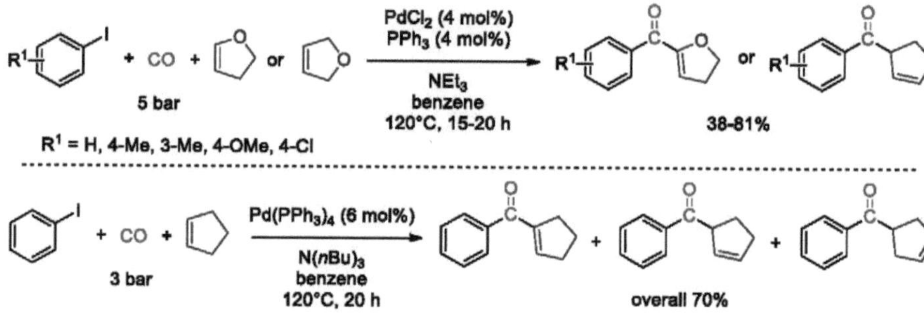

Scheme 5: Carbonylative cross-coupling of aryl iodides with 2,3-dihydrofuranes.

In 1997, Alper and co-workers found allenes to be useful nucleophiles in a palladium-catalyzed carbonylative coupling with o-iodophenols.[43] The reaction proved to be highly regioselective since any asymmetrically substituted allene resulted in only a single isomer of the corresponding benzopyranone product (Scheme 6).

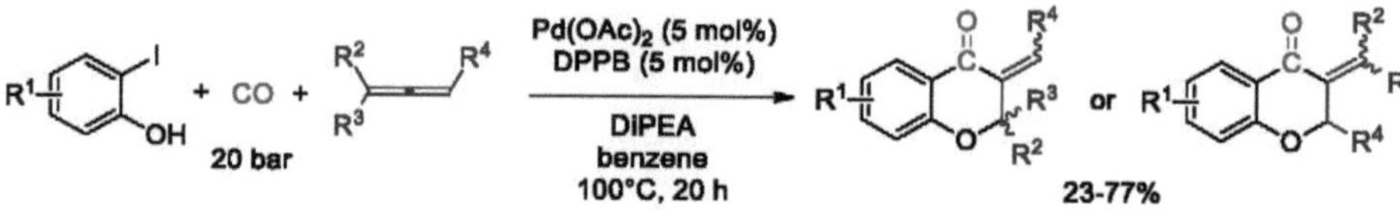

Scheme 6: Carbonylative cross-coupling of 2-iodophenoles with allenes.

A decade later the same group extended their methodology to the analogous carbonylative coupling of o-iodoanilines with allenes.[44] This time the reaction was performed under a decreased CO pressure of 5 bar (Scheme 7). The corresponding 3-methylene-2,3-dihydro-1H-quinolin-4-ones were isolated in yields from 21 to 90%. Notably, the use of an ionic liquid as solvent enabled higher efficiency and allowed for proving the recyclability of the palladium catalyst. Furthermore, it was discussed to be a promoter in the coupling process.

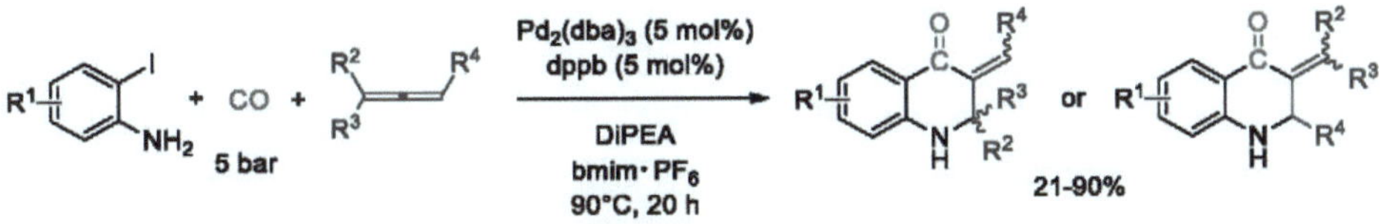

Scheme 7: Carbonylative cross-coupling of 2-iodoanilines with allenes.

Thus far, three-component carbonylative Heck reaction had been limited to fairly active cyclic olefins and allene-type nucleophiles. A first attempt to overcome this limitation was undertaken by the group of Iwasawa in 2007.[45] In a molybdenum carbonyl mediated process, an *in situ* generated acyl-metal species underwent a simple intermolecular addition reaction onto alkenes (Scheme 8). Subsequent protonation released the saturated ketone product while β-hydride elimination and formation of the α,β-unsaturated ketone was not observed.

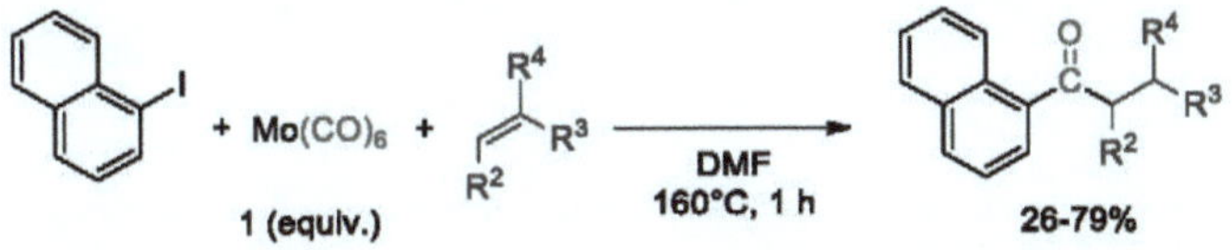

Scheme 8: Addition of *in situ* generated molybdenum-acyl complexes onto alkenes.

In contrast, our group was able to demonstrate a catalytic carbonylative Heck reaction undergoing β-hydride elimination and formation of the α,β-unsaturated ketone. For the first time, readily available styrene derivatives were employed in a palladium-catalyzed carbonylative coupling with aryl and vinyl triflates.[46] The resulting chalcones, also known as natural product flavonoids, are known to have interesting biological activities and represent building blocks for the synthesis of further valuable compounds.[47] The protocol enabled the synthesis of a number of substituted chalcones in moderate to excellent yields and notably, also proved to be applicable to 8 different (internal and external) alkenyl triflates yielding the corresponding dienyl ketones as another class of interesting building blocks (Scheme 9). Unfortunately, the applied catalyst system did not promote the carbonylation of aryl halides.

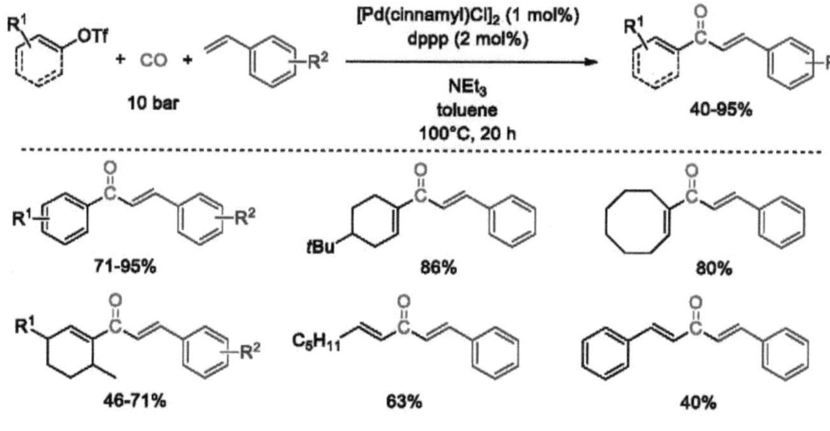

Scheme 9: Carbonylative cross-coupling of aryl/vinyl triflates with styrenes.

Continuing investigations by the same group led to an extension of the palladium catalyzed carbonylative Heck reaction to aryl halides.[48] Hence, aryl iodides were employed successfully in conjunction with aromatic alkenes giving the corresponding α,β-unsaturated ketones in yields ranging from 41 to 90% (Scheme 10). The system was also shown to be suitable for the coupling of aryl bromides with aromatic alkenes. Additionally, it was demonstrated that

2   Carbonylative Coupling Reactions

acrylates and vinyl ethers can be employed when aryl iodides are applied as coupling partners. Key to the success was the application of imidazolium- or pyrrole-based phosphine ligands (**L1** and **L2**) to enable the coupling process of aryl iodides and bromides, respectively. In mechanistic investigations of this catalytic system, the aryl palladium as well as the acyl palladium complex could be characterized by single crystal X-ray analysis, allowing for a detailed investigation of the reaction pathway. In a separate study, this carbonylative Heck coupling was adopted for the synthesis of 3,5-diarylpyrazolines and similar pyrazoles via *in situ* trapping of chalcones with N-methylhydrazine.[49]

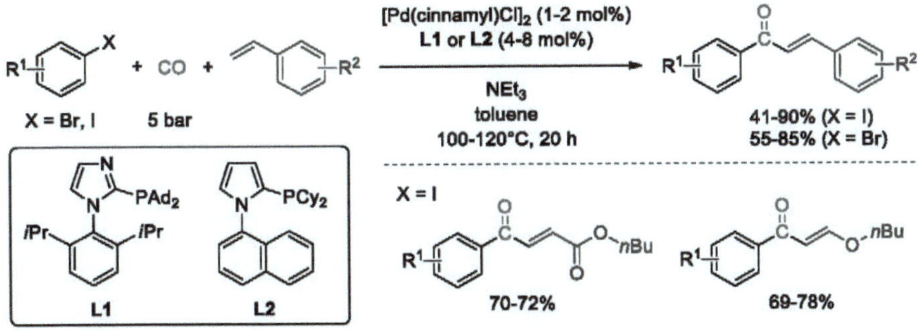

Scheme 10: Carbonylative cross-coupling of aryl halides with olefins.

In a third approach the synthesis of chalcones could be improved further.[50] Following previous work, the synthesis of chalcones from aryl bromides was extended to a broad substrate scope (Scheme 11). 17 different aryl bromides were converted successfully resulting in yields between 45 and 90%, albeit the alkene coupling partner was limited to styrene derivatives. Notably, the carbonylation was achieved with simple PPh$_3$ as a ligand, while DMF was found as most convenient solvent in this case.

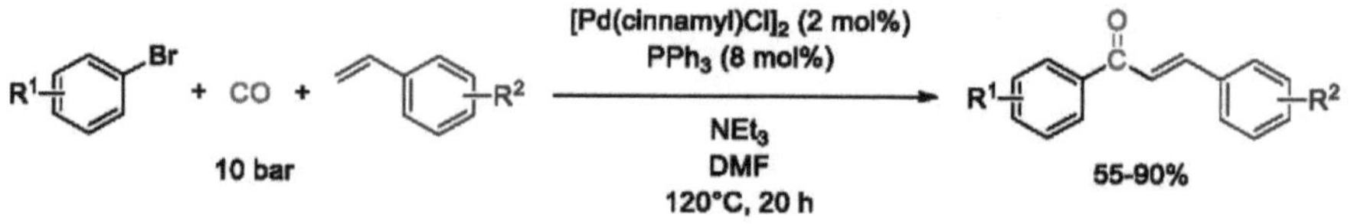

Scheme 11: Carbonylative cross-coupling of aryl bromides with styrenes.

Based on the experimental results, density functional theory computations were then conducted.[50] These calculations confirmed the experimentally observed regioselectivity towards the C–C bond formation on the terminal position of the olefin double bond. Steric influences of the phenyl and acyl groups in the transition states of the C–C bond formation and the β-hydride elimination were found to account for the observed trans-selectivity. Generally, the reaction mechanism is assumed to consist of known elemental steps (Scheme 12) starting with the oxidative addition (6) of ArX to the Pd(0) center to form the corresponding aryl palladium complex. Subsequent coordination (1) and insertion (2) of CO forms the respective acyl palladium complex. Via the following coordination (3), insertion (4) and elimination (5) processes, the desired chalcone is produced. With the assistance of base, the catalytically active Pd(0) species is recycled to enable the next turn over. Regarding DFT calculations, the CO coordination (1) as well as the CO insertion (2) are reversible under the reaction conditions. Therefore, the aryl palladium CO complex and the acyl palladium complex are in equilibrium. Furthermore, the energy barrier for the formation of the direct coupling product (8) is 4.86 kJ/mol higher than that for the formation of the carbonylative coupling product. This small difference in energy confirms the competition of both reaction pathways and the favored formation of the α,β-unsaturated ketone over the formation of stilbene.

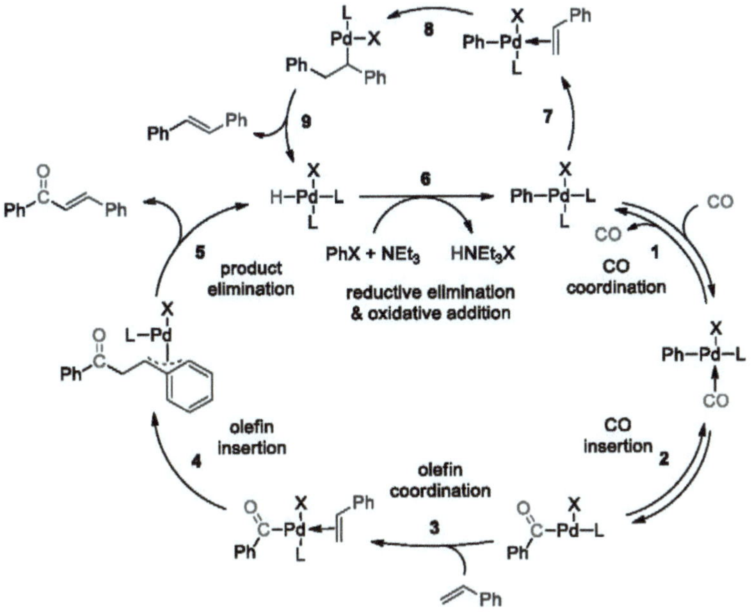

2 Carbonylative Coupling Reactions

Scheme 12: Catalytic cycle for the carbonylative Heck reaction of phenyl halides with styrene.

Instead of using gaseous carbon monoxide in pressure reactors, Skrydstrup and co-workers developed an elegant two-chamber system using a stable, crystalline and non-transition metal based carbon monoxide source COgen (9-methyl-9H-fluorene-9-carbonyl chloride).[51] Employing this intricate CO surrogate, a carbonylative Heck reaction of aryl iodides and styrene derivatives proceeds smoothly and allows for applying near-stoichiometric amounts of carbon monoxide.[52] Moreover, the ^{13}C variant of that CO precursor was prepared and enabled isotope labeling via incorporation of ^{13}CO into the target chalcone product (Scheme 13).

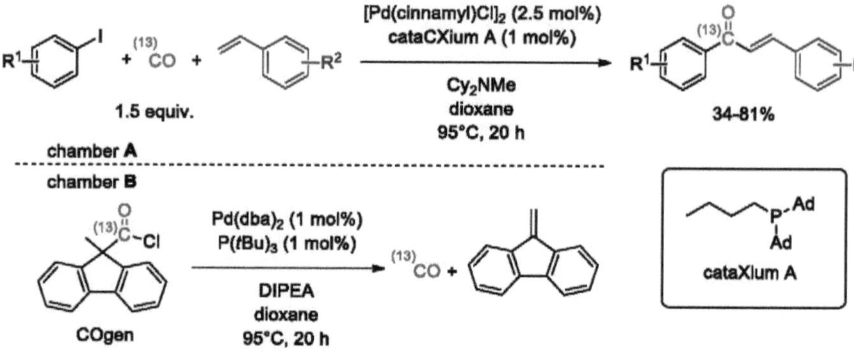

Scheme 13: Carbonylative Heck reaction with styrenes using near-stoichiometric amounts of CO.

While previous studies had focused on the use of aromatic alkenes as nucleophiles and the corresponding synthesis of chalcone derivatives, the application of other functionalized alkenes had rarely been studied. However, in 2012, our group established the first carbonylative Heck coupling reaction of aryl bromides and vinyl ethers.[53] The corresponding results will be discussed in section 4.1 (see page 60). Shortly after, the group of Skrydstrup continued this approach and reported on the carbonylative palladium-catalyzed synthesis of 1-aryl-3-alkoxy propenones starting from aryl iodides.[54] In the presence of [Pd(cinnamyl)Cl]$_2$ and HBF$_4$P(tBu)$_3$, the corresponding products were formed in 41 to 81% yield (Scheme 14). The resulting 1-aryl-3-alkoxy propenones represent valuable monoprotected 1,3-dicarbonyl equivalents, which are known to serve as attractive building blocks for a variety of heterocycles. Once again, their two-chamber system allowed for applying near stoichiometric amounts of carbon monoxide and ^{13}C-labeling of the coupling product and consequently deriving heterocyclic structures.

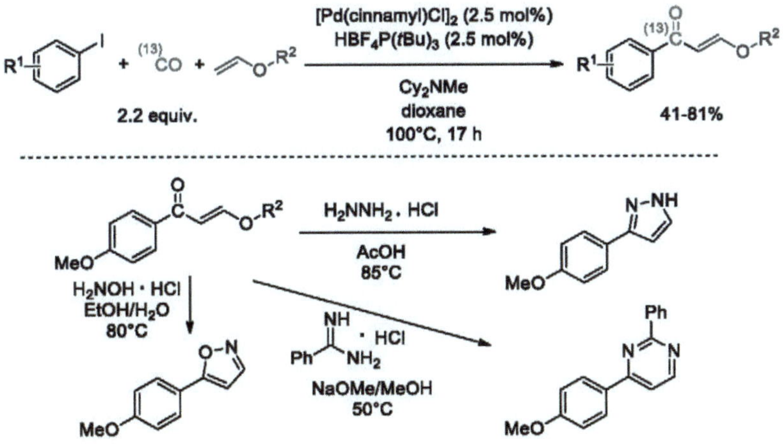

2 Carbonylative Coupling Reactions

Scheme 14: Carbonylative Heck reaction with vinyl ethers using near-stoichiometric amounts of CO.

Based on the palladium-based catalytic system our group had developed for the carbonylative coupling of aryl bromides and vinyl ethers, the reactivity of non-functionalized terminal alkenes was also investigated.[55] Details of this work will be discussed under 4.2 (see page 65). Generally, carbonylative Heck reactions have emerged as a powerful tool for the synthesis of α,β-unsaturated ketones. The application of modern catalytic systems based on palladium and phosphorous ligands has enabled a broad range of substrates with respect to the applicable aryl (pseudo)halides as well as olefins.

2.2 Carbonylative α-Arylations

Complementary to the classical palladium-catalyzed cross-coupling reactions that employ organometallic compounds, alkenes or alkynes as nucleophiles, in the past few decades the α-arylation of C–H acidic compounds as carbon nucleophiles has emerged as a promising methodology. This enables $C(sp^2)$–$C(sp^3)$ coupling through functionalization of C–H bonds. In most cases, sufficient C–H acidity is facilitated by an adjacent carbonyl group and therefore enolate formation plays a primary role in the activation process. In 1997 pioneering work was conducted by Miura, Buchwald, Hartwig and others, that in turn led to multiple synthetic applications of catalytic α-arylation procedures.[19-21, 56] This technology helped to circumvent drawbacks of previously used routes, such as the need for stoichiometric amounts of metal catalysts in conjunction with preformed enolates.[57] Also, the application of unactivated, less-reactive aryl (pseudo)halides had been previously problematic. Allowing for the straightforward functionalization of especially $C(sp^3)$–H bonds, catalytic α-arylation has become a key platform for the development of palladium-catalyzed (hetero)arylation reactions of esters, ketones, nitriles and related α-C–H acidic substrates.[58] Ongoing investigations and mechanistic studies have afforded more efficient catalyst systems and eventually enabled the α-arylation of more demanding, less C–H acidic substrates (i.e. acetone).[59] However, the incorporation of carbon monoxide into the α-arylation process has been studied less extensively, although carbonylative α-arylations can provide easy access to 1,3-dicarbonyl compounds from simple aryl halides and carbonyl derivatives.

In 1986, Kobayashi and Tanaka were the first to conduct a carbonylative α-arylation when they reported on the carbonylative coupling of aryl iodides with activated methylene compounds in the presence of a

PdCl$_2$(dppf) catalyst and base under 20 bar of carbon monoxide.[60] Depending on the nature of the C–H acidic substrate, a variety of different products was observed (Scheme 15). This work concurrently disclosed the potential and difficulties of carbonylative α-arylation reactions. Thus, the conversion of an α-methylated malonate derivative led to carbonylative coupling giving the corresponding α-acyl compound (**A**). Reacting an α-ketoester also resulted in a carbonylative coupling at the α-carbon center, albeit the product was obtained in its enol form (**B**). Diethyl malonate, bearing a CH$_2$-group in between the carbonyl centers, first underwent carbonylative α-arylation followed by an alkoxycarbonylation of the formed enol and resulted in enol ester formation via the eventual incorporation of two molecules of each, CO and Ph (**C**). Finally, direct formation of vinyl esters from the *in situ* formed enolates was observed when the reaction used ethyl acetoacetate, deoxybenzoin or acetophenone as substrates (**D**). The resulting distribution of possible products revealed the challenges of further investigations towards selective and therefore more useful systems for carbonylative coupling reactions of C–H acidic compounds, especially those undergoing enol formation.

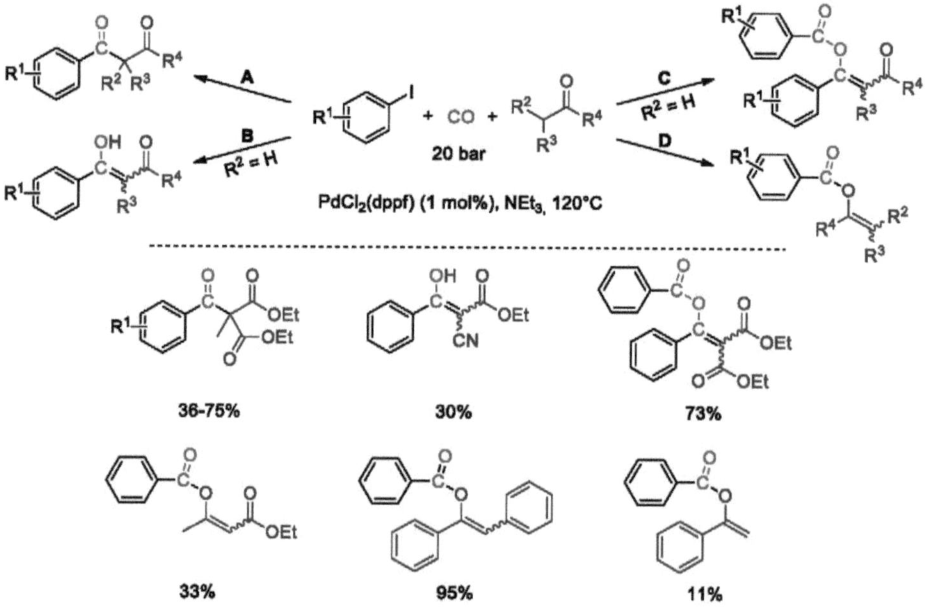

Scheme 15: Carbonylative coupling of aryl iodides with carbonyl derivatives.

A protocol for a more selective carbonylative α-arylation was reported by Negishi and co-workers in 1989.[61] By carrying out the carbonylative coupling in an intramolecular fashion, selectivity problems could be circumvented (Scheme 16). In principle, the reaction can be catalyzed by nickel-, copper- and palladium-based complexes. However, the latter ones proved to be most efficient as well as most general with respect to substrate scope. Thus, this system catalyzed the intramolecular carbonylative cyclization to form 5-, 6- or 7-membered rings in moderate to excellent yields ranging from 56 to 96%. To provide a sufficiently C–H acidic carbon center, doubly stabilized enolates were essential intermediates and attempts to react simple mono ester derivatives (R^3 = H) resulted in no conversion. Hence, malonates, α-nitrile esters, and α-sulfone esters were suitable substrates.

2 Carbonylative Coupling Reactions

Notably, the latter underwent decarboxylation during the carbonylation reaction and no direct non-carbonylative cyclizations were observed under the reaction conditions.

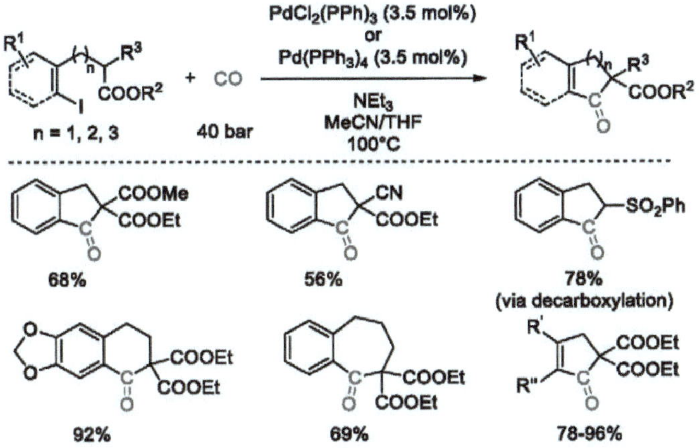

Scheme 16: Intramolecular carbonylative α-arylation/vinylation of activated esters.

A first attempt to conduct a carbonylative α-arylation of ketones was again undertaken by the group of Negishi through an intramolecular reaction.[62] When (2-iodophenyl)acetone was used as a starting material, the 6-O coupling process was favored over the 6-C process, generating the corresponding enol ester product in a 90% yield (Scheme 17). In contrast, 3-(2-iodophenyl)propiophenone underwent 5-C coupling and produced the corresponding cyclic 1,3-diketone in the same high yield. The identical selectivity was observed for a similar 3-vinyl propiophenone derivative. In the case of R = CH_2Ph, the 5-C coupling is favored over 7-C as well as 7-O bond formation, indicating that ring size influences the selectivity in this process. Thus, a selective intermolecular approach remained challenging at this stage.

Notably, the reaction could only be conducted with NEt$_3$ as base, while the application of more basic KOtBu did not lead to any product formation.

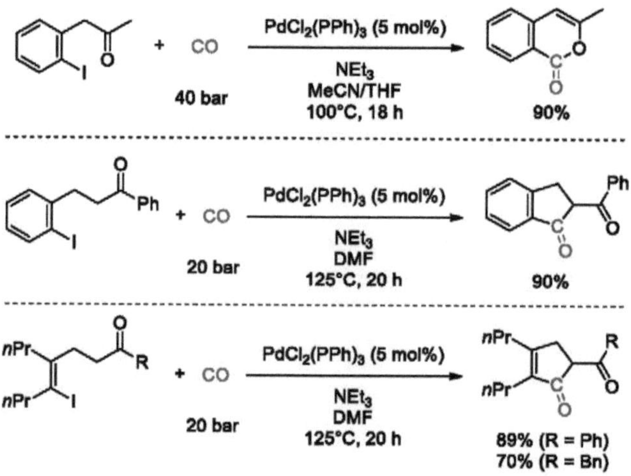

Scheme 17: Intramolecular carbonylative α-arylation of ketones.

In 2007, Wang and co-workers published the first example for the carbonylative coupling of α-diazocarbonyl compounds.[63] The palladium-catalyzed coupling of aryl iodides with ethyl diazoacetate proceeds at a low carbon monoxide pressure (1 bar) and fairly low temperature (45°C) (Scheme 18). While aryl iodides result in the corresponding α-keto-β-diazocarbonyl compounds in moderate yields (43-66%), β-iodostyrene could also be converted into the corresponding ketone in 59% yield. In the same publication, the direct non-carbonylative arylation of α-diazocarbonyl is described in the absence of CO under the same reaction conditions. In comparison, carbonylative coupling showed slightly diminished yields and limited substrate scope, being attributed to decreased reactivity of the catalyst in the presence of carbon monoxide.

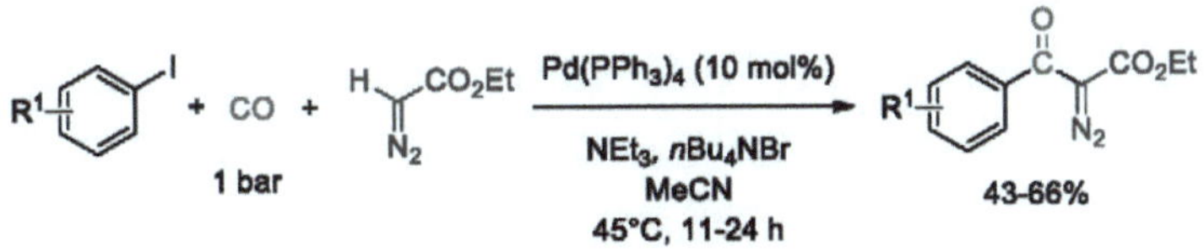

2 Carbonylative Coupling Reactions

Scheme 18: Carbonylative coupling of aryl iodides with ethyl diazoacetate.

A breakthrough in carbonylative α-arylation chemistry was achieved by the group of Skrydstrup in 2012.[64] By reacting aryl iodides with different ketones in the presence of carbon monoxide the intermolecular carbonylative α-arylation of ketones was reported for the first time (Scheme 19). As the catalyst, commercially available Pd(dba)$_2$ in combination with 1,1'-bis(diisopropylphosphino)ferrocene (D*i*PPF) as a bidentate phosphorous ligand proved effective. This approach offers easy access to 1,3-diketones which are known to be valuable precursors to important building blocks and heterocyclic compounds. The reactions were carried out in a two-chamber system (see Scheme 13) that had already been described for other carbonylation reactions and enables the use of near-stoichiometric amounts of carbon monoxide.[52] The ^{13}C-variant of this CO surrogate could again be used to obtain labeled 1,3-diketones and in turn the corresponding pyrazole and isoxazole derivatives after quenching with hydrazine or hydroxylamine. Unlike the systems described before, a strong base (NaHMDS) was required to achieve a sufficient coupling process. It is noteworthy that a polar solvent is required to ensure a high selectivity towards the carbonylated coupling over the direct α-arylation of the *in situ* formed enolate. The authors attribute this effect to the enhanced basicity of the sodium enolate in nonpolar solvents (i.e. toluene).

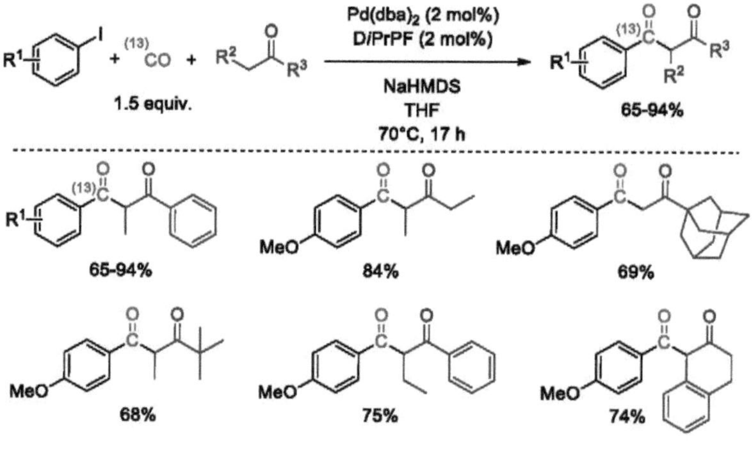

Scheme 19: Intermolecular carbonylative α-arylation of ketones.

In subsequent studies, the same group investigated the carbonylative α-arylation of acetylacetone and similar 1,3-diketones with aryl bromides.[65] Through a palladium catalyzed carbonylative C–C bond formation, the corresponding triketones were obtained (Scheme 20). Key to this success was the use of $MgCl_2$ in combination with NEt_3 forming very reactive enolate surrogates that sufficiently undergo the carbonylative coupling procedure.

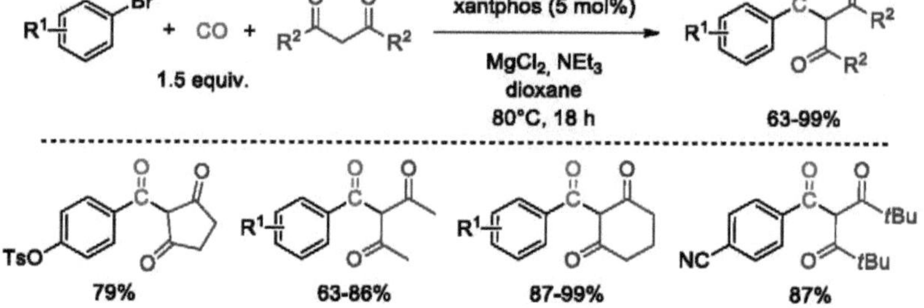

Scheme 20: Carbonylative α-arylation of 1,3-diketones to triketones.

2 Carbonylative Coupling Reactions

More importantly the formed triketones, once treated with aqueous HCl at 80°C, undergo selective deacetylation to form 1-aryl-substituted 1,3-diketones (Scheme 21). The protocol requires only a slight excess of the 1,3-diketone starting materials to generate the corresponding product in moderate to high yields ranging from 44 to 99%. The produced 1-aryl-1,3-butadiones were not accessible via the previously reported carbonylative α-arylation of ketones in the two-chamber system since acetone and cyclohexanone resulted in poor selectivities under those conditions, which led to complex product mixtures derived from competing aldol processes. However, prior to this publication the application of acetone and acetophenone in such a reaction had been reported by our group and will be discussed in chapter 4.4 (see page 73).[66]

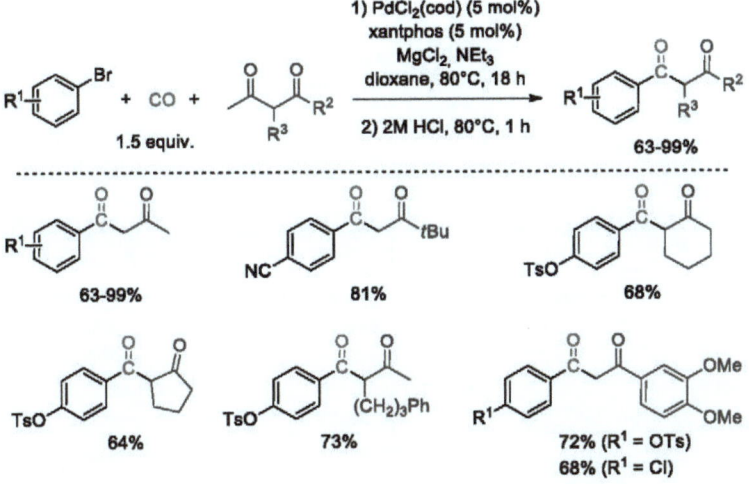

Scheme 21: Carbonylative α-arylation of 1,3-diketones with subsequent deacetylation.

In conjunction with the earlier described protocol, Skrydstrup and co-workers also reported on the employment of monoester potassium malonates as nucleophiles.[67] Aryl bromides were coupled smoothly in the presence of the same catalytic system, assisted by $MgCl_2$ (Scheme 22). In this case, subsequent decarboxylation occurred *in situ* with no need for an additional treatment with HCl, yielding β-ketoesters in yields up to 99%. Notably, the protocol also converts aryl chlorides, triflates and even benzyl chlorides into the corresponding β-ketoesters. Thus applying reactive carboxylic acid chlorides in alternative synthetic routes towards β-ketoesters can be circumvented. Similar to the 1,3-dicarbonyl compounds mentioned before, β-ketoesters also constitute key building blocks in the synthesis of a wide variety of heterocycles that eventually lead to a number of valuable pharmaceuticals and natural products.

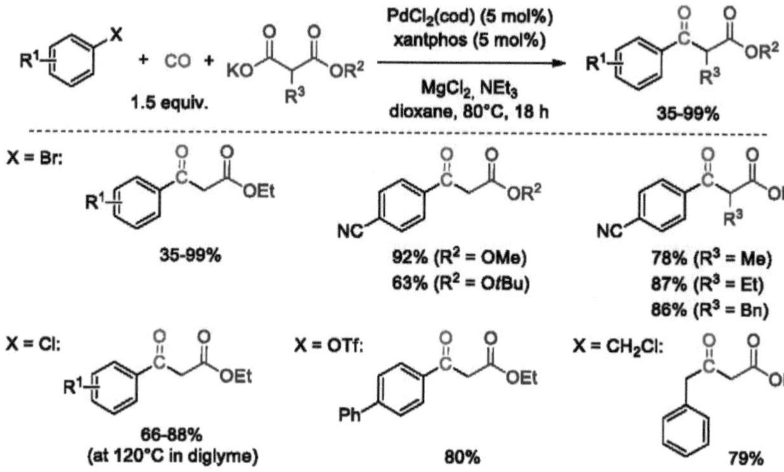

Scheme 22: Carbonylative α-arylation of monoester potassium malonates.

Generally, intermolecular carbonylative α-arylations have gained increasing interest in the past two years. Nevertheless, compared to their non-carbonylative variants they are still less developed, especially with respect to the scope of C–H acidic compounds.

2.3 Carbonylative C–C Bond Formation via C–H Bond Activation of Arenes

In the previous sections, the catalytic functionalization of alkene C–H bonds (carbonylative Heck reaction) and fairly active C–H acidic compounds (carbonylative α-arylation) using aromatic ketones have been discussed. In extension, the direct C–H bond functionalization of arenes constitutes another challenging field in carbonylative C–C cross-coupling chemistry.[68] This approach offers the potential to revise classical catalytic cross-coupling reactions developed thus far, involving the coupling of aryl or vinyl halides with an organometallic reagent (as described before) or other carbon nucleophiles such as alkenes or alkynes.

The direct (non-carbonylative) C–H bond activation and functionalization has been intensively studied over the recent years and has become a valuable tool in organic synthesis.[69] It has been shown that metal C–H bond activation occurs relatively easy forming metal-carbon bonds and cyclometalated species via oxidative addition to electron-rich metal centers or σ-bond metathesis.[70] Even more challenging is the oxidative addition of aryl halides or insertion of unsaturated substrates, which needs to be favored to achieve an efficient catalytic cycle for a C–C bond formation.

Initially, palladium- and rhodium-based complexes have been found to be suitable catalysts for direct C–H bond activation and subsequent C–C bond formation.[71, 72] More recently, cheaper ruthenium catalysts have also proved useful for the efficient conversion of C–H bonds.[73] By combining suitable substrates with effective directing groups a variety of new regioselective cross-coupling procedures has evolved, albeit some challenges still need to be addressed (i.e. lower catalyst loadings, milder reaction conditions and higher functional group tolerance).

2.3.1 Carbonylative C–C Cross Coupling via C–H Bond Activation of Arenes

Until recently, carbonylative C–C coupling reactions via the C–H activation of arenes have only scarcely been studied.[74] This methodology offers easy access to diarylketones, such as benzophenones and derivatives. Typically, the syntheses of such compounds are catalyzed by palladium complexes in carbonylative Suzuki, Negishi, Stille, and Hiyama reactions (see chapter 2, page 23).[32, 34-36] A few reports also make use of organoaluminum and organoindium compounds.[30] A direct C–H bond activation of arenes circumvents the need for organometallic reagents as aromatic carbon nucleophiles and the accompanied production of stoichiometric amounts of waste. Simultaneously, it makes the pre-functionalization of the aromatic carbon nucleophile obsolete.

The first examples of carbonylative cross coupling processes via C–H activation of an arene were published by Campo and Larock in 2000 and 2002 in two individual publications.[39] In a palladium catalyzed cyclocarbonylation of *o*-halobiaryls, various substituted fluorenones were synthesized in high yields (Scheme 23). With catalytic amounts of $Pd(PCy_3)_2$ in conjunction with CsOPiv as base and in the presence of an atmospheric pressure of carbon monoxide, the desired ring formed with full conversion after 7 h at 110°C. The intramolecular approach enabled high selectivity without the necessity for a directing group. Electron-donating as well as -withdrawing substituents were well tolerated, generating the corresponding substituted fluorenones with good regioselectivity and in excellent yields. Furthermore, several heterocyclic starting materials underwent carbonylative ring closure, hence fused isoquinoline, indole, pyrrole, thiophene, benzothiophene, and benzofuran products could be obtained.

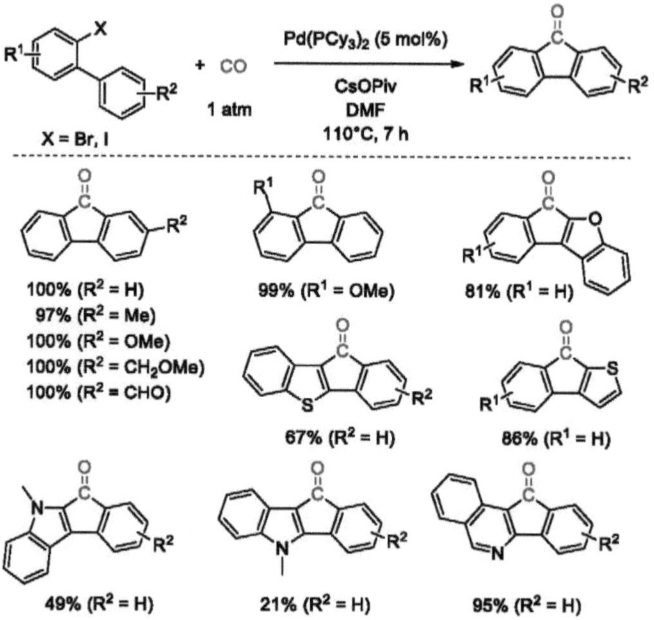

Scheme 23: Intramolecular cyclocarbonylation via C–H activation of arenes.

The first intermolecular example for the carbonylative coupling of aromatic compounds via C–H bond activation was described by the group of Beller in 2010.[41] C–H acidic heteroarenes such as oxazoles, thiazoles, and imidazoles were found to be suitable nucleophiles in a carbonylative coupling reaction with aryl halides (Scheme 24). In this case, a palladium catalyst involving a bidentate ligand (dppp) in combination with stoichiometric amounts of copper(I) iodide was efficient to afford the corresponding diaryl ketones in moderate to very good yields. In the optimization process, the diaryl product deriving from direct coupling was also observed. However, when the CO pressure was increased to 40 bar the desired ketone was obtained exclusively.

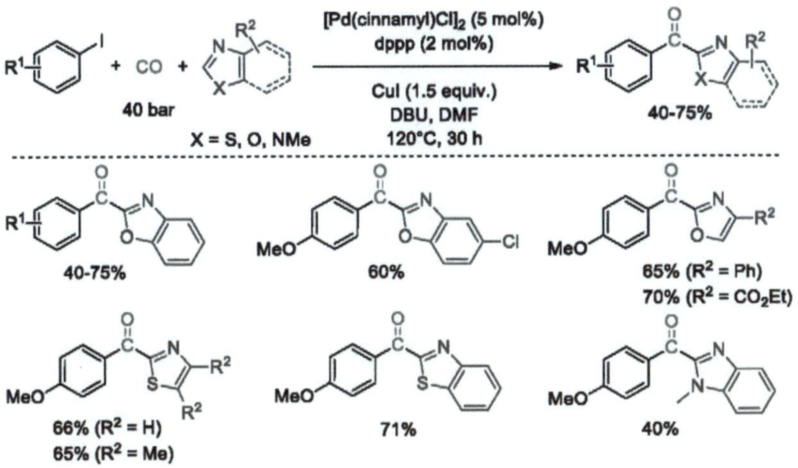

2 Carbonylative Coupling Reactions

Scheme 24: Carbonylative cross-coupling of aryl iodides with heteroarenes.

The mechanism of this transformation is proposed to involve elementary steps known from other well-investigated palladium-catalyzed carbonylative coupling reactions. Initial oxidative addition of iodobenzene onto a palladium(0) species generates an arylpalladium(II) complex that undergoes CO insertion to the corresponding acylpalladium(II) complex. Following transmetallation with the preformed copper-arene species and subsequent reductive elimination enables the product formation and regenerates the active palladium species.

In 2012, the group of Lei reported the palladium-catalyzed oxidative double C–H functionalization/carbonylation of diaryl ethers.[75] As the catalyst Pd(OAc)$_2$ was employed and no additional ligand was necessary (Scheme 25). The reactions were carried out in trifluoroacetic acid (TFA) leading to the *in situ* formation of Pd(TFA)$_2$. This in turn undergoes a fast electrophilic palladation of the diarylether starting material (2). Subsequent CO insertion (3), activation of a second C–H bond (4) and product elimination occur (5).

Finally, $K_2S_2O_8$ acts as an oxidant to recycle the catalytically active Pd(II)-species (1). Mechanistic studies suggest the second C–H bond activation step to be rate-determining. The resulting xanthone derivatives were obtained in yields ranging from 27 to 96%, indicating a significant influence of the nature of the substituents on the cyclization process.

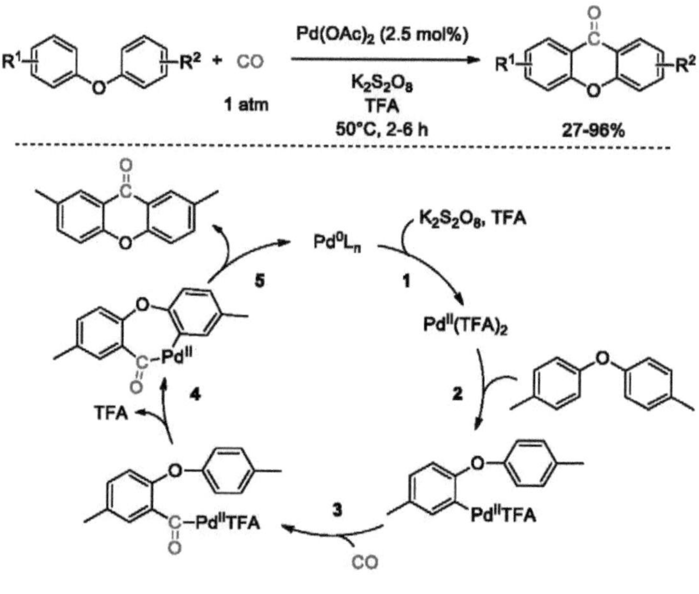

Scheme 25: Intramolecular oxidative double C–H bond functionalization of diarylethers.

Overall, the application of palladium catalysts for the carbonylative C–C cross coupling via C–H activation of arenes is still in the early stages, since as shown here, only a few examples have been described to date.

2.3.2 Carbonylative Hydroarylation via Directing Group Assisted C–H Bond Activation of Arenes

In addition to the classical cross coupling approach, the C–H bond functionalization of arenes has been realized via directing-group assisted C–H bond activation procedures. Ruthenium complexes have proved especially effective in these protocols. This methodology has been extensively studied for forming C–N, C–O and C–C bonds.[76] When amides or alcohols are employed as nucleophilic coupling partners in the presence of an oxidant, these reactions lead to the formation of the corresponding aromatic imides or esters.[77] Although interesting ruthenium-catalyzed hydroarylations were initially developed, related hydroaroylations, leading to ketones by incorporating a molecule of carbon monoxide, have been less studied. Moore and co-workers were responsible for this development in 1992, when they reacted pyridine with 1-hexene in the presence of carbon monoxide to obtain the corresponding ketone.[78] The application of catalytic amounts of $Ru_3(CO)_{12}$ enabled a selective *ortho*-acylation of pyridine and formation of the corresponding saturated ketone in a 65% yield with a 13:1 selectivity towards the linear product (Scheme 26). Mechanistic and kinetic studies suggest the multinuclear feature of the catalyst to be important for promoting the C–H bond activation, involving coordination of the nitrogen of the heterocycle to a ruthenium cluster.

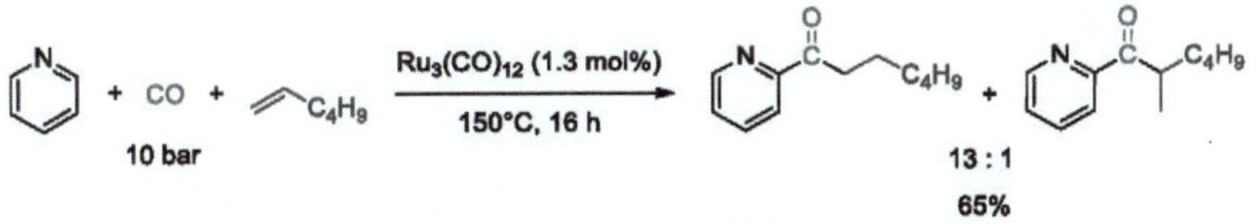

Scheme 26: Reaction of pyridine with carbon monoxide and 1-hexene.

In 1997, the first example of a carbonylative C–H bond functionalization and C–C coupling on a benzene derivative was described by Murai and co-workers.[79] By converting pyridylbenzenes with CO and ethylene in the presence of catalytic amounts of $Ru_3(CO)_{12}$, propionylation occurred selectively on the *ortho* C–H bond of the benzene ring (Scheme 27). In this case, no reaction was observed on the pyridine ring, but it served as directing group to promote the reaction on the benzene ring. In addition to ethylene, trimethylvinylsilane and *tert*-butylethylene could also be employed successfully. In general, rather harsh reaction conditions were required (160°C), for the latter case even higher temperature (180°C) was necessary. For the reaction of 2-phenylpyridine with CO and ethylene, double carbonylation led to a 2,5-disubstituted byproduct, being formed in larger amounts with extended reaction times. Notably, this second functionalization was suppressed if substituents were present at any position of the benzene ring, which gave the corresponding propenones in good yields ranging from 74 to 88%. Also, heterocycles such as thiophene, pyrimidine as well as substituted pyridines could be applied as directing groups. Other than ethylene, 3,3-dimethylbutene and trimethylvinylsilane underwent hydroaroylation, albeit with moderate yields of 34 and 50%, respectively. It is noteworthy to mention that a heterogeneous Ru/C catalyst system was reported to promote the same reaction in 2002.[80]

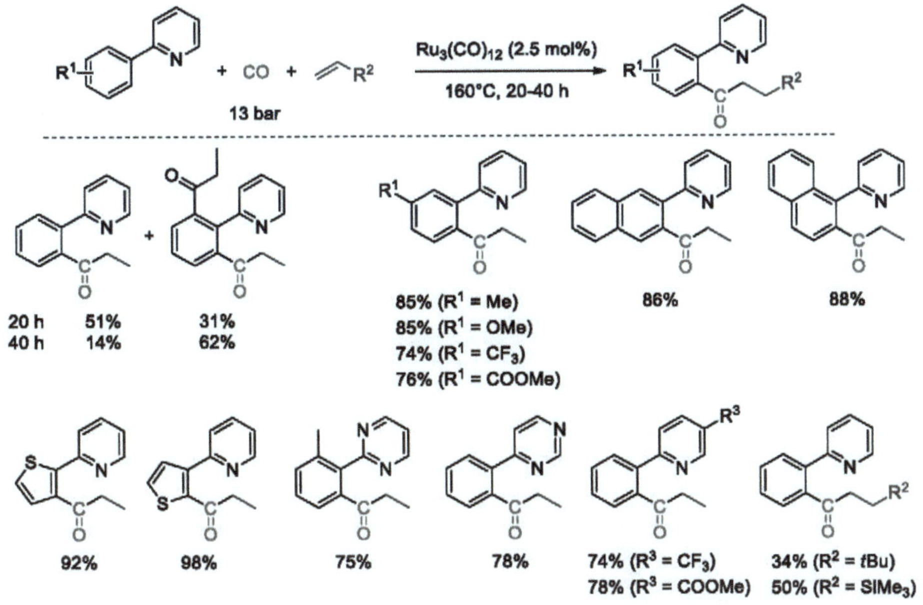

2 Carbonylative Coupling Reactions

Scheme 27: Reaction of pyridylbenzenes with carbon monoxide and ethylene.

The catalytic cycle has not fully been elucidated yet, although individual ruthenium species have been characterized successfully.[81] Initially, the coordination of the ruthenium to the sp^2-nitrogen atom in the pyridine directs the metal complex to the *ortho*-position of the arene (Scheme 28). Subsequent C–H bond cleavage gives the five-membered ruthenacycle (**1**). Following coordination (**2**) and hydrometalation (**3**) of ethylene generates an ethyl ruthenium complex. Carbon monoxide then inserts into either the C(aryl)-Ru or C(Et)-Ru bond to form the corresponding Acyl-Ru complex (**4**). Finally, the product is released via reductive elimination and the catalytically active ruthenium species is regenerated (**5**).

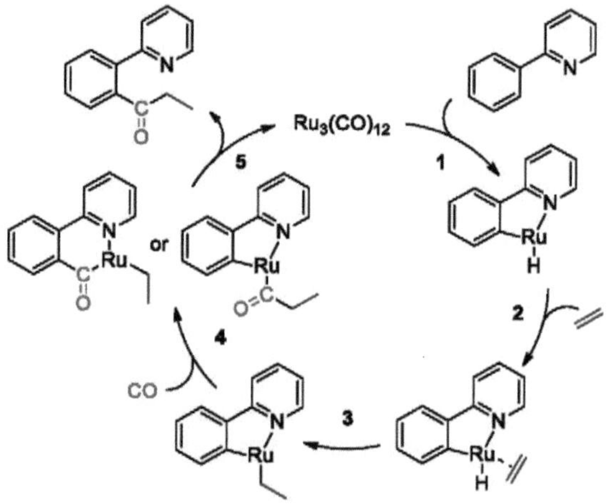

Scheme 28: Proposed catalytic cycle for the reaction of 2-pyridylbenzene with CO and ethylene.

In 1998, the group of Murai extended this methodology towards the site-selective carbonylative C–H functionalization of aza-heterocycles.[82] With the same catalytic system, 1,2-disubstituted benzimidazole and 2-methylimidazo[1,2-α]pyridine were smoothly converted into the corresponding aryl alkyl ketones (Scheme 29). Notably, the C–H bond functionalization occurred selectively in the β-position of the unsubstituted nitrogen in the five-membered ring. In contrast to the reaction of 2-phenylpyridine, 1-hexene, 2-methylstyrene and vinyl cyclohexene could also be applied as alkenes to yield the corresponding ketones in yields between 25 and 78%.

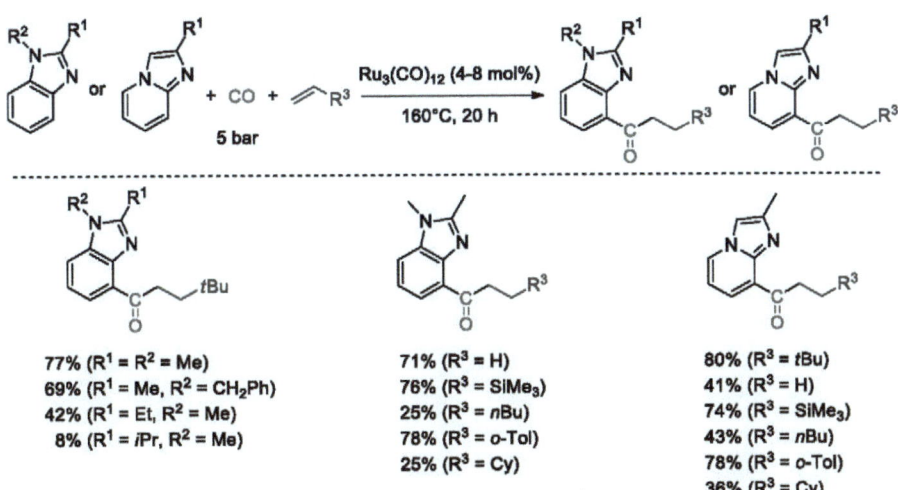

Scheme 29: Reaction of aza-heterocycles with carbon monoxide and alkenes.

Ongoing investigations by the same group revealed the applicability of oxazoline and pyrazole as directing groups in the carbonylative hydroarylation of alkenes via C–H bond activation of arenes.[83] When oxazoline served as a directing group, mono- as well as disubstitution was observed in the *ortho*-position of the arene moiety (Scheme 30). This accounts for the large variation of the corresponding yields (42 to 98%), since disubstitution was blocked in the case of *ortho*-substituted starting materials, in turn favoring the formation of the mono-ketone product. In contrast, monosubstitution occurred exclusively in cases where N-aryl pryridine was applied as the starting material. Interestingly, oxazine and thiazoline also proved effective as directing groups in the ruthenium catalyzed hydroaroylation of ethylene, although the scope of the olefin was scarcely investigated in these reports.

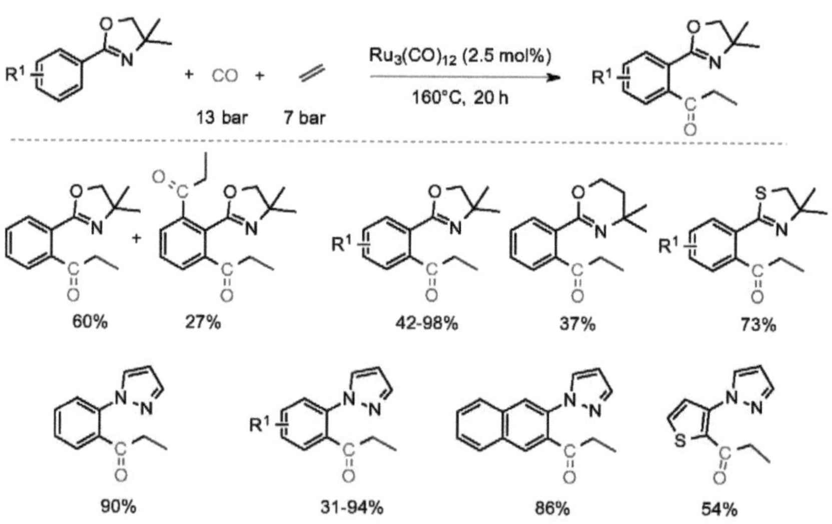

Scheme 30: Oxazoline and pyridine acting as directing groups.

Finally, N-pyridylindolines have been described as another compound class that is suitable for a directing-group assisted carbonylative addition to ethylene.[84] In 2002, Chatani et al. proved the principle by reacting 5 different N-pyridylindoline derivatives with carbon monoxide and ethylene (Scheme 31). In this case, the influence of the substituent on the directing group was investigated, showing that methyl substitution can be beneficial or hindering depending on its position on the pyridine ring. Attempts to apply propylene as the starting material resulted in mixtures of the linear and branched aryl alkyl ketone in a low overall yield of 25%.

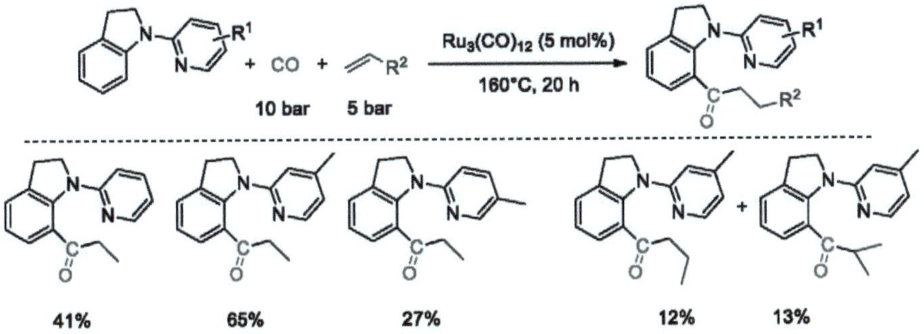

2 Carbonylative Coupling Reactions

Scheme 31: Reaction of *N*-pyridylindolines with carbon monoxide and alkenes.

As discussed thus far, several different directing groups have proven to be effective in the carbonylative C–C bond formation via C–H bond activation of arenes. However, the coupling partners were limited to a few olefins, such as active ethylene. In comparison, analogous non-carbonylative C–H bond functionalization procedures feature a variety of alkenes, alkynes as well as aryl, vinyl, and benzyl (pseudo)halides, all of which have been found to be reactive in the C–C bond formation.[76]

3   Objectives of This Work

Deriving from the consistent progress in catalyst and ligand design, transition metal-catalyzed reactions have become a powerful tool in synthetic organic chemistry. Due to high demand on the catalyst, especially carbonylative cross-coupling can benefit from the recent developments. Generally, carbonylative C–C bond formation offers easy access to a variety of valuable products. The incorporation of carbon monoxide into the target molecule does not only utilize a cheap and abundant C1 building block but also simultaneously installs useful functionality. In the last decade, most of the well-known coupling reactions have been extended to their carbonylative variants. However, the majority of these reactions require the employment of stoichiometric amounts of an organometallic reagent to deliver a sufficiently nucleophilic carbon center to the coupling process.

In conjunction with recent efforts towards more sustainable chemical procedures, more efficient syntheses and waste reduction have become important objectives. In general, three-component coupling reactions can enhance efficiency in comparison to alternative multi-step syntheses of the same products. With regards to carbonylative coupling reactions of activated arenes with carbon nucleophiles, a first step towards an increased sustainability is constituted in the direct functionalization of a C–H bond and the adherent abandonment of metal organic reagents.

Inspired by the direct Sonogashira and Heck reactions, alkynes and alkenes have been found to be reactive nucleophiles for forming the corresponding α,β-unsaturated ketones in three-component coupling reactions as well. Especially the latter reaction has been studied less extensively with respect to the scope of functionalized alkenes. However, the employment of different (functionalized) olefins is crucial to make the

3 Objectives of This Work

carbonylative Heck coupling a versatile and broadly applicable process. Therefore, an extension of the known systems beyond styrene derivatives has been an objective of this work.

Another approach exhibits the carbonylative α-arylation of C–H acidic compounds. Again inspired by the non-carbonylative version, such reactions enable the direct synthesis of 1,3-difunctionalized compounds. Depending on the group initiating sufficient acidity of the α-C–H bond, an entire toolbox of coupling procedures can be developed. When the research for this thesis was started, no example for an intermolecular carbonylative α-arylation of unactivated ketones or nitriles had been reported. Thus, the conducted research aimed for the development of the corresponding three-component coupling procedures.

To achieve another step towards more sustainable C–C bond formation procedures, the carbonylative coupling via C–H bond activation of non-activated arenes represented an additional goal. Deriving from first reports on directing-group assisted procedures catalyzed by ruthenium complexes, further developments towards the implementation of this technique into carbonylative C–C coupling reactions with different coupling partners was desirable.

Finally, within this thesis, there has been a persistent interest in the development of other new transition metal-catalyzed reactions utilizing carbon monoxide.

4 Summary of Results

4.1 Carbonylative Heck Reaction of Aryl Bromides with Vinyl Ethers

Arising from the direct Heck reaction leading to substituted alkenes, as well as from the Heck carbonylation reactions generating carboxylic acid derivatives, carbonylative Heck-type coupling reactions to α,β-unsaturated ketones have been developed. Corresponding comprehensive studies have been reported by our group as well as by Skrydstrup and co-workers.[46, 48-51]

While previous studies focused on the use of aromatic alkenes as nucleophiles and the corresponding synthesis of chalcone derivatives, the application of other functionalized alkenes had rarely been studied and carbonylative coupling of aryl bromides with vinyl ethers failed under the previously described conditions. Since the resulting 1-aryl-3-alkoxy-propenones represent valuable mono-protected 1,3-dicarbonyl equivalents, which are known to serve as attractive building blocks for a variety of heterocycles,[85] it appeared worthwhile to investigate their synthesis in more detail.[53]

Initially, the reaction of bromobenzene with carbon monoxide and an excess amount of n-butyl vinyl ether was investigated as model system (Scheme 32). Applying $PdBr_2$/**L2** as the catalyst system generated the desired product in a 60% yield after optimization of the reaction conditions, including solvent and base. Some more commercially available ligands were then tested. Notably, several 2-dicyclo-hexylphosphinopyrroles turned out to be capable ligands for the model system. While cataCXium POMeCy (**L4**) gave the best results (76% yield), the structurally similar SPhos ligand (**L6**)

4 Summary of Results

showed only low activity under the same conditions. Therefore the electronic properties of dicyclohexylphosphine substituted pyrrole ligands as well as the sterical demand caused by substituion (**L3**) of the pyrrole fragment or by either *N*-naphthyle (**L2**) or *ortho*-substituted *N*-aryl-moieties (**L4, L5**) seem to be crucial.

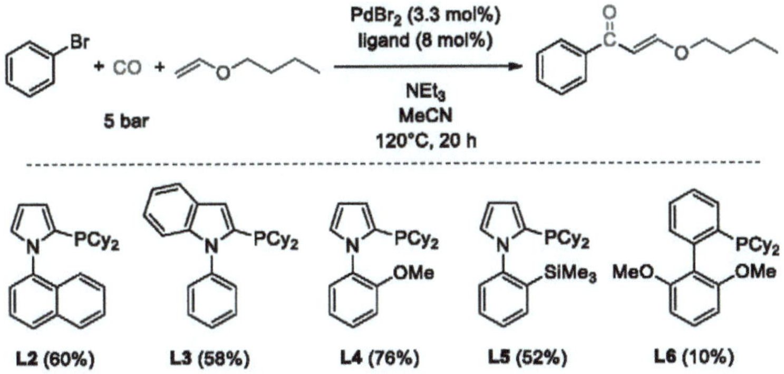

Scheme 32: Variation of ligands in the carbonylative Heck reaction of vinyl ethers.

Interestingly, the reaction proceeded in a highly regioselective manner and no products incorporating external double bonds were detected. Furthermore, the *trans*-coupling product was detected exclusively, while the Heck coupling product was observed in less than 5% yield.

The generality of the protocol was demonstrated in the carbonylative coupling of 16 different aryl bromides with *n*-butyl vinyl ether (Scheme 33). Substituents in *para*-, *meta*- and even *ortho*-positions on the aryl ring were well-tolerated and resulted in moderate to very good yields. Electron-rich and slightly electron-deficient aryl bromides led to formation of the desired product in yields ranging from 50 to 70%. It is noteworthy, that more electron-poor aryl bromides such as 4-bromobenzonitrile, -acetophenone, and -benzaldehyde underwent the coupling process with decreased yields. However, 2-bromonaphthalene and several sulfur and nitrogen containing heteroaromatic substrates were converted smoothly into the corresponding 1-aryl-3-butoxy-2-(E)-propen-1-ones with moderate yields (56 to 70%).

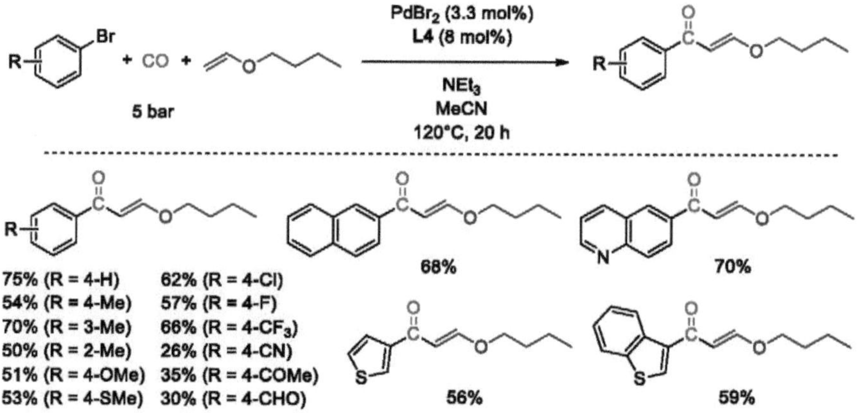

Scheme 33: Variation of aryl bromides in the carbonylative Heck reaction with vinyl ethers.

In further experiments, the applicability of the catalytic system on the carbonylative vinylation with different vinyl ethers was investigated (Scheme 34). Several alkyl vinyl ethers were successfully converted into the corresponding 1-phenyl-3-alkoxy-propenones in yields from 51 to 73%. Notably, employing styrene under the optimized conditions revealed an enhanced performance, resulting in the formation of the corresponding chalcone in an excellent 99% yield.

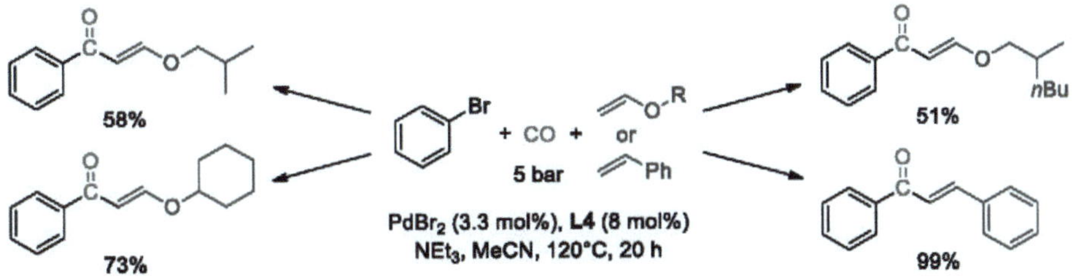

4   Summary of Results

Scheme 34: Variation of vinyl ethers in the carbonylative Heck reaction with aryl bromides.

The resulting 1-aryl-3-alkoxy propenones represent valuable monoprotected 1,3-dicarbonyl equivalents, which are known to serve as building blocks for a variety of heterocycles. Hence, the coupling process, primarily giving 1-aryl-3-alkoxy propenones could also be implemented into a one-pot synthesis of pyrazoles, which are known to have a broad spectrum of biological activities.[86] Through this novel approach, aryl-substituted pyrazoles were synthesized in moderate to good yields (Scheme 35).

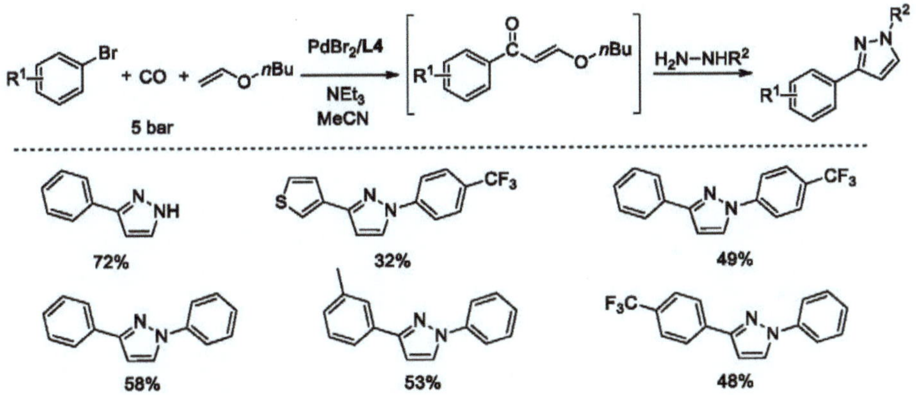

Scheme 35: One-pot synthesis of pyrazoles.

In conclusion, the first carbonylative Heck coupling reaction of aryl bromides and vinyl ethers has been established. Allowing for the employment of vinyl ethers and representing an enhanced system for carbonylative coupling of styrenes, the novel protocol is a valuable extension of the recently published coupling of aryl halides with styrenes.[48, 48-51] 19 different 1-aryl-3-alkoxy-2-propen-1-ones were synthesized in a straightforward manner. Based on this coupling process a new and easy route to 1,3-diaryl-pyrazoles has been developed demonstrating the usefulness of the synthesized products as building blocks. Thus, several pyrazole derivatives were prepared efficiently from simple bromobenzene, CO, vinyl ether and phenyl hydrazine. It is noteworthy that a very similar protocol for the carbonylative coupling of aryl iodides with vinyl ethers was published by the group of Skrydstrup following to this report.[54] For further details, see publication 5.1, *Chem. Eur. J.* **2012**, *18*, 4827–4831.

4   Summary of Results

4.2 Doublecarbonylation Reaction of Aryl Halides to Arylfuranones

Based on the catalytic system our group had developed for the carbonylative coupling of aryl bromides and vinyl ethers, the reactivity of non-functionalized terminal alkenes was investigated.[55] Surprisingly, significant amounts of products incorporating two molecules of carbon monoxide were observed that could be identified as 5-arylfuranones. These furanones, also referred to as α,β-unsaturated butenolides, constitute a structural motif shared by many biologically interesting natural compounds that show anti-leukemic,[87] anti-malarial,[88] anti-tumor[89] or anti-inflammatory properties.[90] Considering the availability of the corresponding starting materials, a direct synthesis of furanones from (hetero)aryl halides, alkenes and carbon monoxide would represent a valuable extension of the known procedures.

Initially, when three different phenyl (pseudo)halides were reacted with 1-octene in the presence of CO, palladium catalyst and base, four different coupling products were observed (Scheme 36). The use of phenyl triflate as the starting material exclusively led to the formation of α,β-unsaturated ketones I and II with the linear *trans*-product being predominant. While employing iodobenzene resulted in mono- and dicarbonylated products, the application of bromobenzene led to a remarkable selectivity towards the incorporation of a second CO molecule. Thus, in the latter case, the furanone IV was formed in a 59% yield, whereas only low amounts of the corresponding monocarbonylation product II and only traces of compounds I and III were detected.

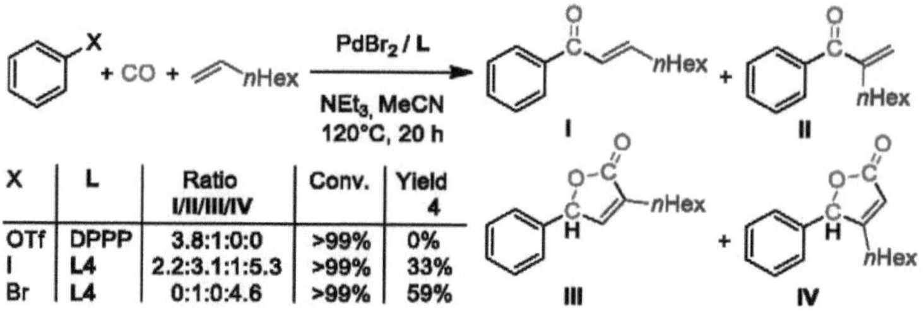

4 Summary of Results

X	L	Ratio I/II/III/IV	Conv.	Yield 4
OTf	DPPP	3.8:1:0:0	>99%	0%
I	L4	2.2:3.1:1:5.3	>99%	33%
Br	L4	0:1:0:4.6	>99%	59%

Scheme 36: Carbonylative Heck coupling of PhOTf, PhI and PhBr with 1-octene.

Due to the good selectivity towards the formation of 4-*n*-hexyl-5-phenylfuran-2(5*H*)-one (**IV**) the carbonylative coupling of bromobenzene with 1-octene was applied as model system for further investigations. An important detail in the optimization process was raising the CO pressure to 80 bar that in turn enhanced the product formation to a 79% yield under the optimized conditions. While the use of bases was limited to NEt$_3$, several palladium precursors besides PdBr$_2$, such as [Pd(cinnamyl)Cl]$_2$ promoted the reaction similarly well. Also different pyrrole-based cataCXium P ligands proved active, although cataCXium POMeCy (**L4**) remained most efficient.

Next, the scope of aryl bromide was investigated. Methyl-, *tert*-butyl-, methoxy- and fluorine-substituted aryl bromides were converted into the corresponding 4-*n*-hexyl-5-arylfuran-2(5*H*)-ones in yields ranging from 55 to 83% (Scheme 37). Unfortunately, in cases of stronger electron-withdrawing substituents on the aryl bromide moiety, only traces of the corresponding products were observed. However, 2-bromonaphthalene as a bicyclic substrate and also heterocyclic compounds were transformed with up to 87% yield. Notably, the optimized conditions also allowed for the selective conversion of aryl iodides into the desired 5-arylfuranones. At a decreased reaction temperature of 100°C, different substituted aryl iodides were converted in moderate to very good yields (49-87%).

4 Summary of Results

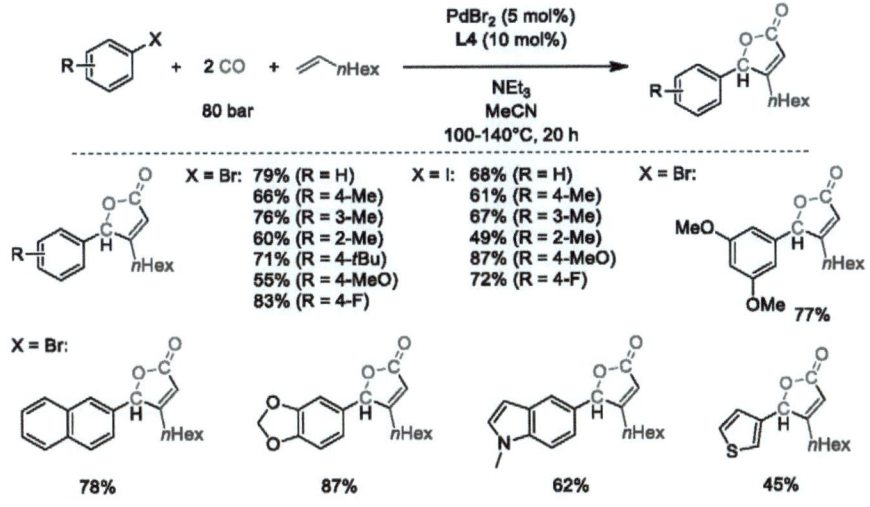

Scheme 37: Palladium-catalyzed doublecarbonylative coupling of aryl iodides and bromides with 1-octene.

When the possibility to employ different terminal alkenes was examined, a slight decrease in yield was observed for substrates with a lower boiling point such as 1-hexene (Scheme 38). The more bulky vinylcyclohexane resulted in a 54% yield, but triethylvinylsilane was converted into the corresponding furanone with a good yield of 73%. 4-Phenyl-1-butene reacted with a decreased selectivity generating the corresponding product in only 28% yield, while also elevated amounts of the isomeric form (according to III) were observed.

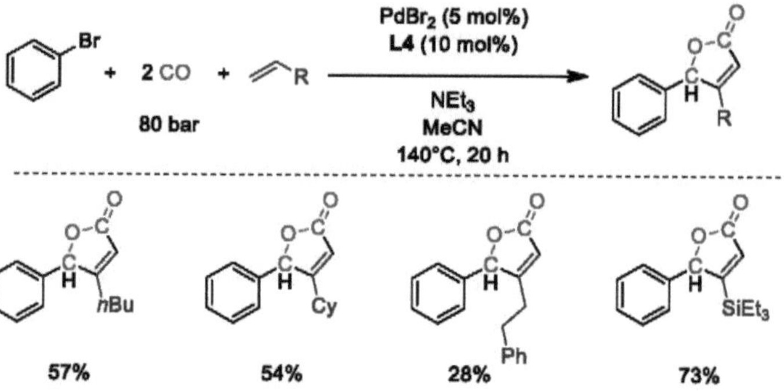

Scheme 38: Palladium-catalyzed doublecarbonylative coupling of bromobenzene with terminal alkenes.

To explain the unusual formation of **IV**, a possible reaction mechanism has been proposed (Scheme 39). As known from studies on the carbonylative Heck coupling of styrenes, initial oxidative addition (**1**) and then CO insertion (**2**) result in the corresponding acyl palladium species. Next, coordination and insertion (**3**) of 1-octene on the acyl palladium center will occur, followed by a base induced enolate formation (**4**). Surprisingly, the intermediate with the branched alkyl chain is favored. Apparently, a subsequent insertion of a second molecule of CO (**5**) is faster than ß-hydride elimination. This can be explained by an intramolecular coordination of the carbonyl group or the respective enolate, in turn preventing β-hydride elimination. Finally, an intramolecular attack of the generated enolate (**6**) forms the product with concomitant regeneration of the active palladium(0) complex.

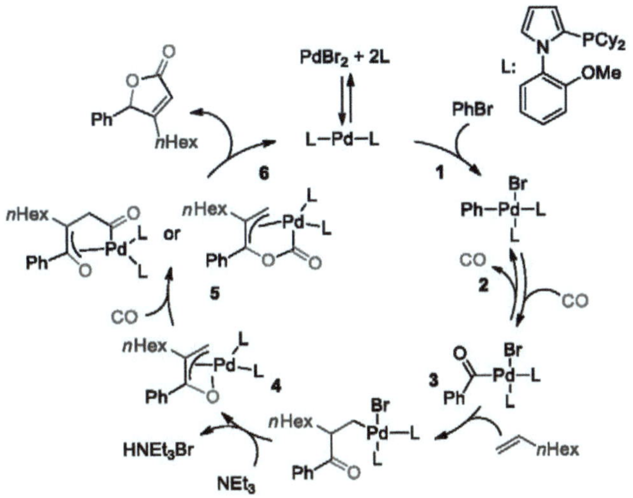

4 Summary of Results

Scheme 39: Proposed mechanism for the doublecarbonylative coupling of bromobenzene with 1-octene.

The described reaction herein represents the first selective synthesis of 5-arylfuranones from aryl halides, CO and terminal alkenes. The incorporation of two molecules of carbon monoxide into the desired product has enabled the highly regioselective production of 18 different 5-arylfuranones in up to 83% yield. This represents a valuable extension of the known procedures yielding 4-alkyl-5-aryl-substituted furanones,[91] especially since most of the isolated compounds have been synthesized for the very first time. For further details, see publication 5.2, *Chem. Eur. J.* **2013**, *19*, 12959–12964.

4.3 Carbonylative Arylation of Aryl Ketones to Vinylbenzoates

Inspired by the palladium-catalyzed direct (hetero)arylation of esters, ketones and related C–H acidic substrates, the analogous carbonylative coupling reactions have stimulated interest recently. Since the pioneering work of the groups of Miura, Buchwald, and Hartwig, such reactions have become popular for the formation of new C(sp2)–C(sp3) bonds, thereby generating allylic and benzylic carbonyl derivatives.[19-21, 56-59] In 2012, Skrydstrup and co-workers described the palladium-catalyzed carbonylative α-arylation of ketones to form 1,3-diketones.[64] In contrast to such C–C bond-forming procedures, related carbonylative C–O bond formations between ketones and aryl halides have been scarcely described.[60, 62] Since vinyl acetates represent valuable substrates in a variety of reactions[92] and the use of vinyl benzoates in contrast has hardly been investigated to date,[93] the carbonylative coupling of C–H acidic compounds that results in the corresponding vinyl benzoates provides an interesting option.

After initial studies had shown that corresponding O-benzoylated compounds were formed when (hetero)aryl halides were reacted with specific ketones in the presence of carbon monoxide, the development of a general methodology for this type of reaction was initiated. Thus, the reaction of iodobenzene with deoxybenzoin under 10 bar of CO was studied in the presence of [Pd(cinnamyl)Cl]$_2$ and nBuPAd$_2$ (cataCXium A) as the ligand. During the optimization process, the role of base was found to be crucial for activating the ketone, and among the several bases tested, Cs$_2$CO$_3$ gave the best result. Notably, the combination of several 2D NMR techniques revealed the selective formation of (Z)-1,2-diphenylvinyl benzoate. Further variations of the reaction conditions such as decreasing the substrate concentration and

4 Summary of Results

temperature optimization finally resulted in full conversion and a 96% yield (determined by GC) of the desired product.

Subsequently, different substituted aryl iodides as well as iodothiophenes were employed successfully in this transformation on a preparative scale (Scheme 40). Generally, electron-donating substituents were better tolerated than electron poor ones, hence the yield varied between 44 and 88%. To extend the scope of the developed coupling methodology, aryl bromides were also tested in the transformation to vinyl benzoates. At 120°C in THF as solvent, sufficient reactivity was observed and the corresponding products were isolated in yields ranging from 49 to 73%.

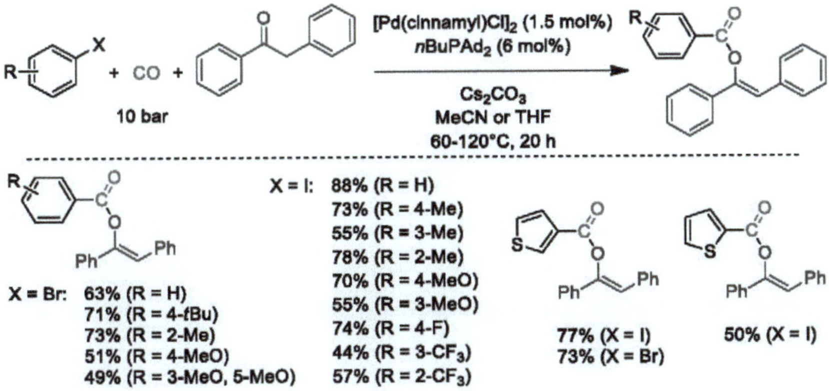

Scheme 40: Palladium-catalyzed carbonylative coupling of aryl bromides and iodides with deoxybenzoin.

Next, the scope of ketone coupling partners was examined (Scheme 41). Using 1-phenyl-2-*p*-tolylethanone, the corresponding product was isolated in a 75% yield. In addition, benzoylation of chloro- as well as bromo-substituted deoxybenzoin derivatives proceeded smoothly and resulted in 66 and 52% yield, respectively. Noteworthy was that 1,3-diphenyl-2-propanone gave the corresponding product in a high yield of 86%. While simple aliphatic ketones did not react under the optimized conditions, the more activated 2-methoxy-1-phenylethanone could be applied successfully.

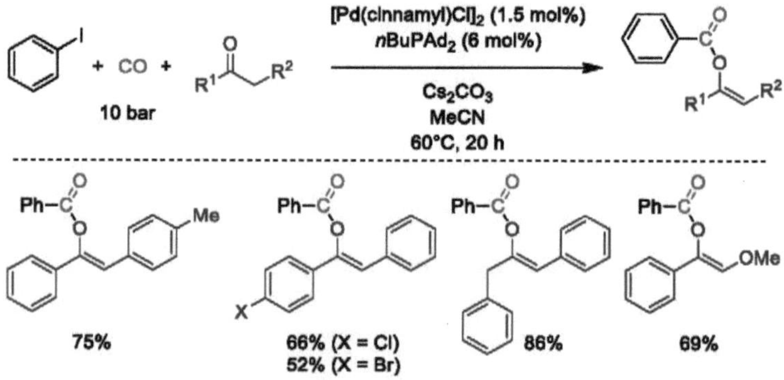

Scheme 41: Palladium-catalyzed carbonylative coupling of iodobenzene with different aryl ketones.

In conclusion, a general carbonylative coupling of aryl halides with benzylic ketones has been developed, that exclusively forms vinyl benzoates. In contrast to previous protocols dealing with palladium-enolate species generated from benzylic ketones, a selective C–O bond formation occurs. The applied catalytic system enables a high regioselectivity and allowed for the first-time synthesis of several vinyl benzoates. For further details, see publication 5.3, *Chem. Eur. J.* **2012**, *18*, 15592–15597.

4.4 Carbonylative α-Arylation of Acetone and Acetophenones

Inspired by the carbonylative arylation of aryl ketones to vinylbenzoates described earlier on (see chapter 4.3), related carbonylative coupling reactions of aryl halides and ketones via C–C bond formation became interesting. As mentioned before, Skrydstrup and co-workers reported the first system that allows for the intermolecular carbonylative α-arylation of ketones, thereby preventing the alkoxycarbonylation of the enolate derivative to an acylated enol.[64] Their system nicely demonstrates the direct synthesis of 1,3-diketones from aryl iodides and simple ketones, although the protocol was not suitable for the most simple and abundant ketones like acetone and acetophenone. However, the direct α-arylation of these more demanding, less C–H acidic compounds has been described initially by the group of Stradiotto, and later by Ackermann as well as Ma and their co-workers.[59]

Obviously, the carbonylative α-arylation of acetone and other easily available ketones is important since the resulting 1,3-diketones are commonly featured in numerous biologically relevant compounds and are valuable building blocks for a plethora of heterocyclic moieties.[94] Hence, the direct synthesis of 1,3-diketones from aryl halides, carbon monoxide and acetone as a simple and abundant carbon feedstock would represent a desirable alternative to the known procedures, most often requiring two reaction steps such as aldol condensation of enolates and carbonyl compounds and subsequent oxidation of the resulting β-hydroxy ketones.

Previously, it was reported, that the [Pd(cinnamyl)Cl]$_2$/nBuPAd$_2$ catalyst system promotes the carbonylative coupling of aryl halides with deoxybenzoin derivatives to the corresponding vinylbenzoate compounds.[95] Surprisingly, the same catalyst system enabled the selective formation of the corresponding 1,3-diketone when less C–H acidic acetone was employed.

Initially, iodobenzene was reacted with acetone under 10 bar of CO in the presence of [Pd(cinnamyl)Cl]$_2$/nBuPAd$_2$ and Cs$_2$CO$_3$ as base, and 1-phenylbutane-1,3-dione was obtained in a promising 19% yield. After an increased reaction time of 20 h had been found to be beneficial to the yield, testing different solvents did not show any positive effect. Also variations in the palladium precursor or the applied ligand did not result in any improved performance. However, the use of 2 ml of acetone as the solvent improved the yield dramatically to 52%, thus the use of additional solvent could be omitted. Also, when several other inorganic (K$_2$CO$_3$, K$_3$PO$_4$, KOH, CsF) or organic (KHMDS, NaOtBu, KOtBu) bases were employed in the model conditions, the desired 1,3-diketone was formed in only low yields. While 3 bar of carbon monoxide was found to be the optimal pressure for the transformation (67% yield), it was demonstrated that even an atmospheric pressure of CO enables the formation of the desired product in a 62% yield. Finally, a further extension of the reaction time to 36 h enabled nearly full conversion and a 71% yield of the desired diketone, representing the optimized reaction conditions.

Then we turned our interest to the scope of the aryl iodide coupling partner (Scheme 42). Similar to the model reaction, 3- and 4-iodotoluene were converted into the corresponding products in good yields of 68 and 71%, respectively. Although 3,5-dimethyliodobenzene and 4-*tert*-butyliodobenzene gave only slightly lower yields (54 and 62%, respectively), alkoxy substituents on the aryl iodide substrates proved to be detrimental to the coupling process and 4-fluoroiodobenzene resulted in an enhanced yield (74%); the relationship between the nature of the substituent on the aryl ring and the yield is unclear. However, bicyclic as well as heterocyclic (pyridyl,

4   Summary of Results

thiophenyl) aryl iodides turned out to be more challenging as the desired 1,3-diketones were obtained in decreased yields between 41 and 56%.

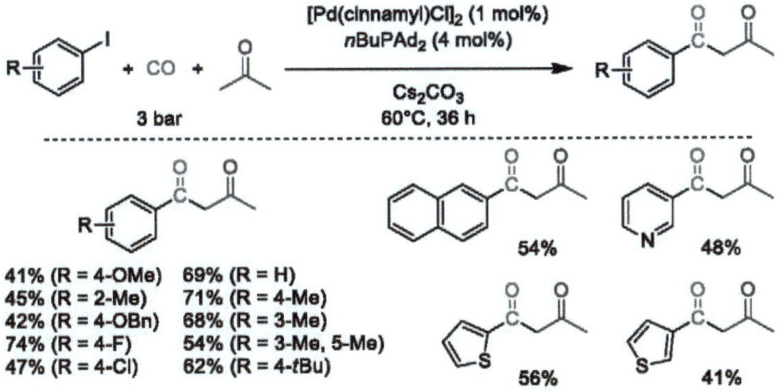

Scheme 42: Palladium-catalyzed carbonylative coupling of acetone with different aryl iodides.

Reaction: R–Ar–I + CO (3 bar) + acetone → 1,3-diketone
Conditions: [Pd(cinnamyl)Cl]$_2$ (1 mol%), nBuPAd$_2$ (4 mol%), Cs$_2$CO$_3$, 60 °C, 36 h

Yields:
- 41% (R = 4-OMe)
- 45% (R = 2-Me)
- 42% (R = 4-OBn)
- 74% (R = 4-F)
- 47% (R = 4-Cl)
- 69% (R = H)
- 71% (R = 4-Me)
- 68% (R = 3-Me)
- 54% (R = 3-Me, 5-Me)
- 62% (R = 4-tBu)
- 2-naphthyl: 54%
- 3-pyridyl: 48%
- 2-thienyl: 56%
- 3-thienyl: 41%

Further studies focused on the employment of different ketones (Scheme 43). Starting the reaction from aliphatic 2-butanone yielded the corresponding product in a 73% yield. Aromatic ketones such as acetophenone and related derivatives were also well tolerated. However, stronger electron-withdrawing groups as well as *ortho*-substitution on the acetophenone moiety turned out to be more challenging in this system. Notably, 2-acetylthiophene as a representative heterocyclic substrate underwent the coupling smoothly, resulting in a 42% yield.

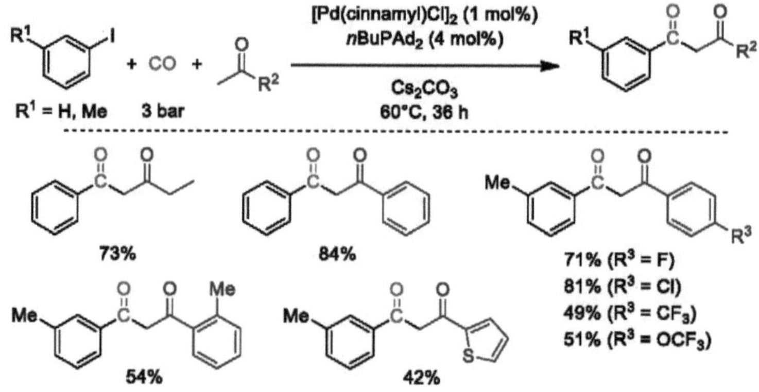

Scheme 43: Palladium-catalyzed carbonylative α-arylation of different ketones to 1,3-diketones.

To prove the usefulness of the developed coupling reaction for the synthesis of building blocks and in turn heterocyclic compounds, a one-pot synthesis of aryl-substituted pyrazoles was assembled (Scheme 44). Pyrazole derivatives are known to have a broad spectrum of biological activities.[86] A known route towards their synthesis is the cyclo-condensation of hydrazines with 1,3-dicarbonyl compounds.[96] Hence, after the palladium-catalyzed coupling process was finished, the remaining acetone was removed *in vacuo*. A subsequent condensation with an aqueous hydrazine solution generated the corresponding 5-aryl-3-methyl-1*H*-pyrazoles in acceptable overall yields from 51 to 62%.

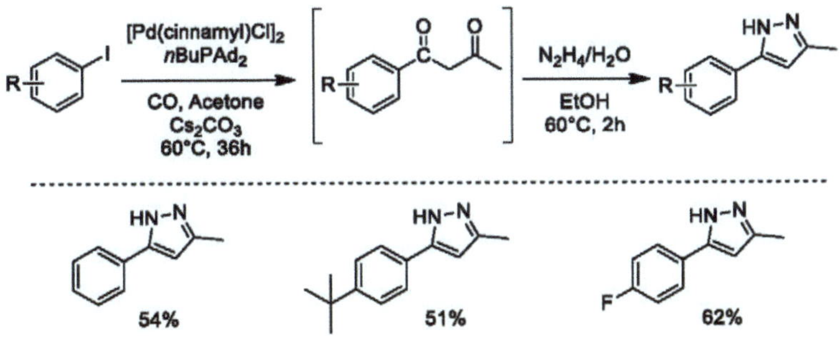

4 Summary of Results

Scheme 44: One-pot synthesis of pyrazoles via carbonylative α-arylation of ketones.

In conclusion, the described system represents the first carbonylative α-arylation of acetone as well as acetophenones. 23 different 1,3-diketones could be synthesized in a straightforward manner in yields ranging from 41 to 84%. Notably, in all optimization reactions no side products derived either from the non-carbonylative α-arylation or multiple arylations of acetone were detected. The isolated 1,3-diketones were isolated as the ketoenol tautomer. Furthermore, atom-economical gaseous CO was applied for the first time in an intermolecular carbonylative α-arylation reaction of ketones. Therefore, a valuable extension of the recently published carbonylative α-arylation requiring the prior synthesis of a CO surrogate has been achieved. The employment of a simple and commercially available catalyst system and the abandonment of additional solvent make the protocol easily applicable. The value of the produced 1,3-diketones as building blocks has been proven by a one-pot synthesis of 5-aryl-3-methyl-1H-pyrazoles. For further details, see publication 5.4, *Chem. Eur. J.* **2013**, *19*, 12624–12628.

4.5 Carbonylative α-Arylation of Nitriles

As described vide supra, carbonylative coupling reactions between aryl halides and C–H acidic compounds are of current interest in homogeneous catalysis. So far, these reactions have been mainly limited to carbonyl-containing compounds that easily undergo enolate formation. In order to extend this methodology, the carbonylative α-arylation of nitriles constitutes a worthwhile goal for further developments. Starting from aryl halides, carbon monoxide and nitriles, all of them being easily available coupling partners, β-ketonitriles can be accessed directly. The resulting β-ketonitriles are useful difunctional intermediates, especially in the preparation of γ-amino alcohols.[97] Apparently, previous synthetic routes have been limited to the formation of β-ketonitriles unsubstituted at the α-position, which in turn has a significant impact on the range of structures that can be accessed.[98]

So far, the direct α-arylation of nitriles has been reported in only a few protocols by Hartwig, Verkade and their co-workers.[25, 99] Hence, initial experiments were inspired by the catalytic system that had been successfully reported for the direct α-arylation of nitriles by the group of Hartwig.[25] Unfortunately, first attempts to apply $Pd(OAc)_2$/BINAP as catalyst for the coupling of bromobenzene and isobutyronitrile in the presence of 5 bar of carbon monoxide did not lead to any formation of the desired β-ketonitrile and only direct coupling was observed. However, starting the reaction from iodobenzene instead and applying an elevated CO-pressure of 30 bar enabled formation of 2,2-dimethyl-3-oxo-3-phenylpropanenitrile for the first time in 21% yield. In further optimizations of the model system, it was shown that only strong bases provide sufficient deprotonation of the nitrile to enable the coupling process, thus NaHMDS gave the best results. When the influence of different ligands was tested, next to the initially used BINAP, also

4 Summary of Results

4,5-bis(diphenylphosphino)-9,9-dimethylxanthene (Xantphos) also promoted the reaction, even in an enhanced conversion and yield. However, a lack of selectivity had induced the formation of two coupling products, the carbonylative and the direct non-carbonylative one. This issue could be solved by applying milder reaction conditions (80°C, 5 bar of CO), a higher catalyst loading and most importantly a slight excess of the ligand (5 mol%) with respect to the palladium precursor (4 mol%). Hence, a high selectivity towards the formation of the β-ketonitrile was achieved, resulting in the desired product in a good isolated yield of 71%.

The optimized reaction conditions were then applied to different aryl iodide substrates (Scheme 45). In doing so, *ortho*, *meta* and *para* mono- as well as dialkyl-substitution were well-tolerated resulting in good to excellent product yields from 75 to 87%. Moreover, alkoxy substituents on the aryl iodide proved to be marginally detrimental to the coupling process and resulted in yields ranging from 69 to 78%. However, chloro- and fluoro-substitution on the iodobenzene moiety also generated the corresponding β-ketonitrile in high yields (71 to 87%) Electron deficient trifluoromethyl substitution was also tolerated, albeit leading to similarly diminished yields as the methoxy substituent and therefore the relationship between the nature of the substituent on the aryl ring and the yield remains unclear. Additionally, bicyclic 1-iodonaphthalene and heterocyclic aryl iodides were employed successfully. Notably, attempts to apply other aryl bromides resulted in consistently low selectivities towards the carbonylated coupling product whereas aryl chlorides led to low conversions.

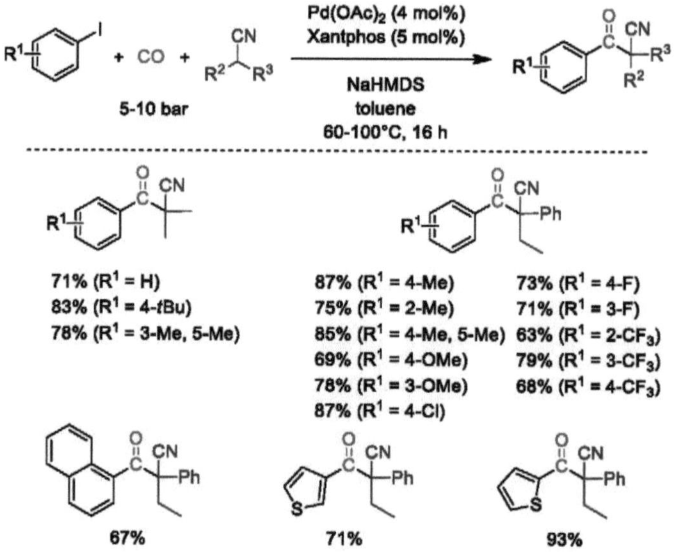

Scheme 45: Palladium-catalyzed carbonylative coupling of nitriles with different aryl iodides.

Furthermore, the scope of the nitrile coupling partner was investigated (Scheme 46). Besides isobutyronitrile, 2-methylbutyronitrile was also transformed into the desired β-ketonitrile in a satisfactory yield of 83%. Additionally, the aroylation of α-substituted benzyl cyanide substrates proceeded smoothly generating the corresponding products in yields of 72 to 86%. Nitriles having a CH_2-group in the α-position were not successful under these conditions since subsequent aroylation of the products in their enol form occurred leading to the corresponding enol benzoates.

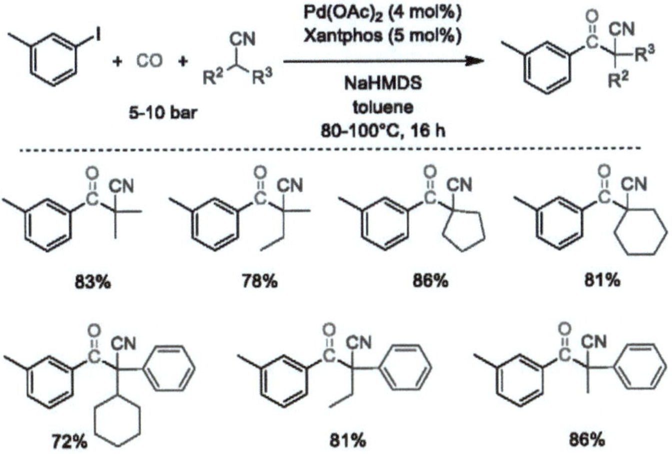

4 Summary of Results

Scheme 46: Palladium-catalyzed carbonylative coupling of 3-iodotoluene with different nitriles.

Next, it was possible to perform the described carbonylation reaction in the two-chamber system developed by Skrydstrup and co-workers (see chapter 2.1, Scheme 13).[51] Since the *ex situ* generation of CO from 9-methyl-9*H*-fluorene-9-carbonyl chloride (COgen) allows for the use of stoichiometric amounts of carbon monoxide, it became reasonable to conduct some labeling experiments (Scheme 47). Thus, ^{13}CO was generated from the corresponding ^{13}COgen in one chamber and fed to the coupling process in a second connected chamber. Consequently, starting the reaction from either iodobenzene or 1-chloro-4-iodobenzene, the ^{13}C-labeled β-ketonitriles were obtained in yields of 50 and 45%, respectively.

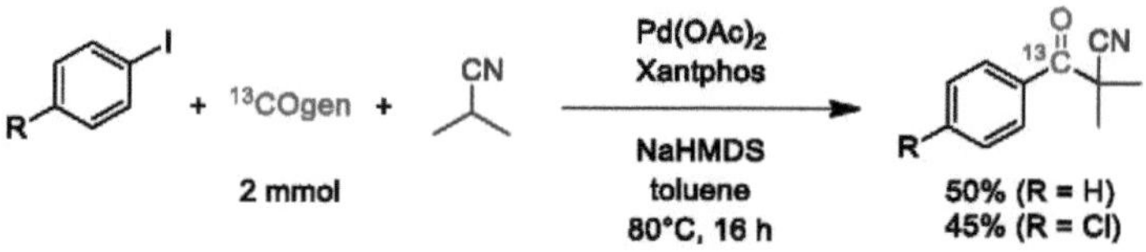

Scheme 47: Carbonylative α-arylation of isobutyronitrile with ^{13}COgen.

In order to prove the usefulness of the synthesized β-ketonitriles as building blocks, a metal-free selective catalytic hydrosilylation procedure was applied that has been recently reported by our group.[100] Hence, a β-ketonitrile obtained from the coupling process reacted smoothly with phenyl silane in the presence of a catalytic amount of TBAF at room temperature (Scheme 48). Notably, a selective reduction of the carbonyl group was observed whereas the nitrile moiety remained untouched. After subsequent hydrolysis the corresponding α-hydroxynitrile could be isolated in a very good yield of 88%.

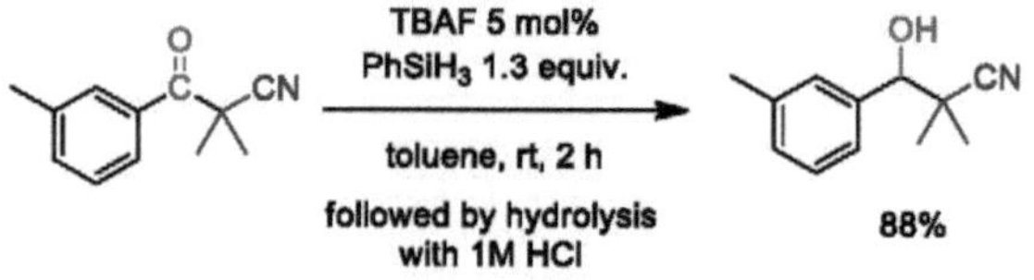

Scheme 48: Selective reduction of β-ketonitrile to β-hydroxynitrile.

In conclusion, the first example of a carbonylative α-arylation of nitriles has been reported. The developed protocol was shown to exhibit a broad substrate scope as 24 different β-ketonitriles were isolated in good to excellent yields. Also, the catalytic system proved versatile allowing for the employment of either atom-economical gaseous CO or a CO surrogate. The latter is capable of ^{13}C-labeling of the generated carbonyl center. Additionally, the usefulness of β-ketonitriles as building blocks has been demonstrated in a straightforward synthesis of β-hydroxynitriles. This in turn emphasizes the resulting β-ketonitriles to be useful difunctional intermediates, i.e. for the synthesis of many biologically active compounds.[101] For further details, see publication 5.5, submitted manuscript.

4.6 Carbonylative Transformations of Bromhexine

Deriving from a variety of carbonylative coupling reactions that have been developed by our group within the last decade, interest aroused to apply these technologies to a pharmaceutically active compound. Hence, bromhexine was found to be a suitable candidate for these tests since the molecule bears two aromatic bromide moieties that qualify for further functionalization.[102] This drug is frequently used in the treatment of respiratory diseases, such as acute and chronic bronchitis.[103] Clinical studies have shown that bromhexine derivatives show a significantly improved impact on the clearing of mucus and also increase the production of pulmonary surfactant.[104] Additionally, bromhexine and its derivatives have been found to act as inhibitors of the human glucocerebrosidase enzyme and therefore can potentially attenuate the course of Gaucher disease.[105] Inspired by the biological properties of bromhexine derivatives, the application of some of the developed protocols was initiated, in order to gain access to new biologically active compounds.

In this context, bromhexine was first applied in a carbonylative Suzuki-Miura coupling with aryl boronic acids. Based on a previously reported system,[34] Pd(OAc)$_2$/nBuPAd$_2$ was found as suitable catalyst system in the presence of 10 bar of carbon monoxide and TMEDA as base. Thus, a variety of arylboronic acids could be employed successfully, yielding predominantly the dicarbonylated product (Scheme 49).

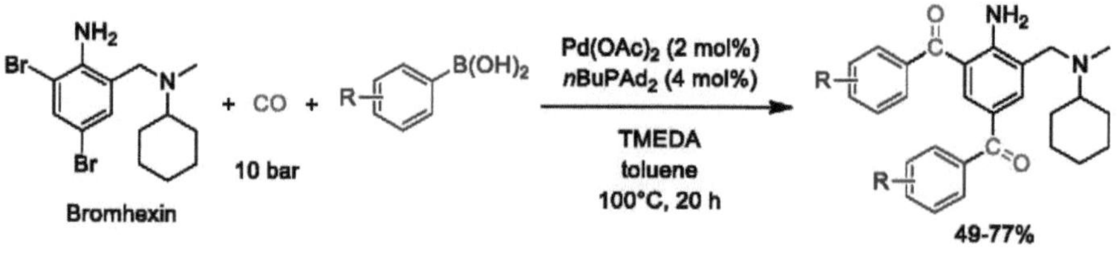

Scheme 49: Palladium-catalyzed carbonylative coupling of bromhexine with arylboronic acids.

Next, the alkoxycarbonylation and aminocarbonylation of bromhexine towards the corresponding esters and amides was investigated. Notably, the same catalytic system allowed for the coupling of several alcohols and amines with bromhexine in the presence of carbon monoxide (Scheme 50). Aliphatic alcohols as well as primary and secondary amines reacted smoothly to the corresponding esters and amides resulting in moderate to good yields (39 to 72%).

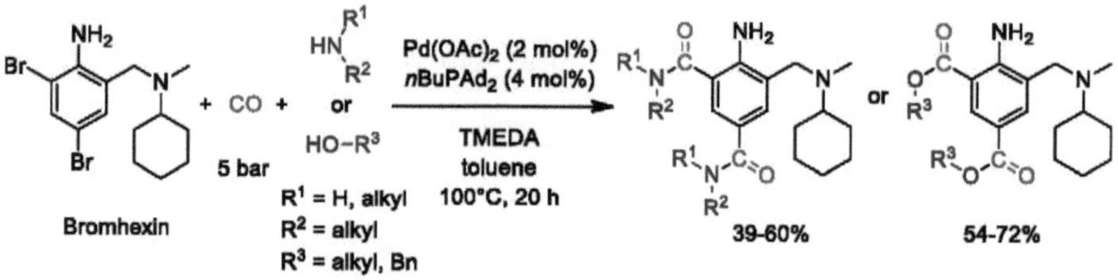

Scheme 50: Palladium-catalyzed carbonylative coupling of bromhexine with alcohols and amines.

4   Summary of Results

To examine the biological function of the synthesized compounds, a subset of the derivatives was subjected to an in vitro GCase inhibition assay (Scheme 51).[105] Notably, compound **2** blocked glucocerebrosidase activity to a similar extent as bromhexine and ambroxol, whereas compound **1** proved to be a significantly more potent inhibitor.

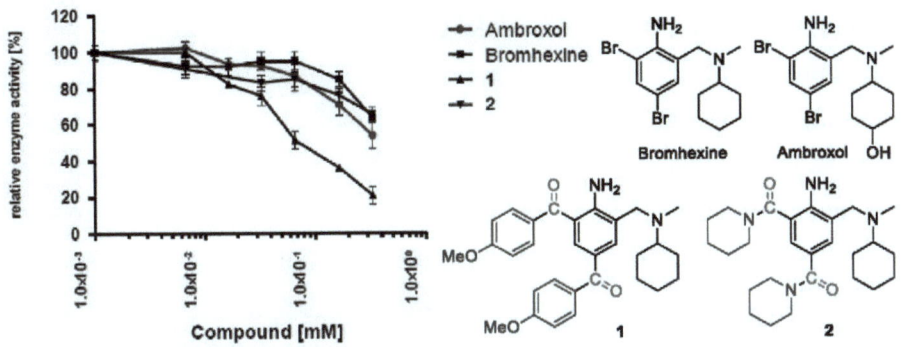

Scheme 51: Inhibition of recombinant GCase by bromhexine and chosen compounds.

Overall, a general synthesis for dicarbonylated derivatives of the parent drug bromhexine has been developed by using a commercially available catalyst system. Preliminary tests have proven biological activity of individually synthesized compounds and represent likely candidates for further pharmacological tests. For further details, see publication 5.6, *Eur. J. Org. Chem.* **2014**, 222–230.

4.7 Aroylation of Aryl Iodides via Directed C–H Bond Activation

Among the palladium-catalyzed coupling reactions of aryl halides, carbonylative Suzuki, Negishi, Stille and Hiyama reactions represent useful tools to access benzophenone derivatives.[32, 34-36] These benzophenones constitute attractive targets, especially those with heterocyclic substituents since they show a plethora of biological activities.[106] To overcome the necessity for the use of stoichiometric amounts of organometallic coupling reagents that is comprehended by the known protocols, it became interesting to conduct a direct C–H bond functionalization of the arene to be coupled with an aryl halide in a carbonylative fashion. Several groups have reported on the C(sp2)–C(sp2) bond formation between arenes bearing a directing group and aryl halides.[76] However, carbonylative C–H bond functionalizations have been less studied and the carbonylative coupling reaction of aryl halides via directed C–H bond activation has not been reported to date.

As a model reaction, 2-phenylpyridine was reacted with phenyl iodide and carbon monoxide (30 bar) in the presence KOAc as additive, K_2CO_3 as base and $[Ru(cod)Cl_2]_n$ polymer as the catalyst. When different additives as well as organic solvents were tested, only minor conversion occurred and the desired carbonylative product was only observed in low yields. In compliance with previous reports,[107] carrying out the reaction in water enabled good reactivity and in turn led to the desired benzophenone derivative in a 56% yield. Further optimizations, especially the employment of sodium bicarbonate as base allowed for good product formation resulting in a 65% yield.

By applying the previously optimized conditions the substrate scope was investigated with respect to the employment of different aryl iodides (Scheme 52). Hence, *ortho*-, *meta*-, and *para*-alkyl substituted aryl iodides

4 Summary of Results

could successfully be converted into the desired products in yields between 52 and 74%. Surprisingly, *ortho*-methyl substitution resulted in the highest yield of 74%. Iodoanisole derivatives as well as electron-poor aryl iodides were slightly less reactive, thus the corresponding benzophenone derivatives were isolated in yields between 41 and 54%.

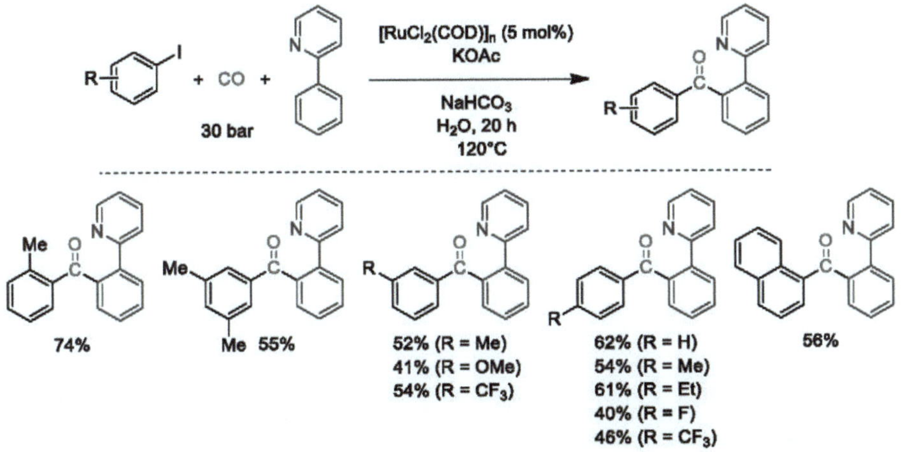

Scheme 52: Ruthenium-catalyzed carbonylative C–H functionalization with different aryl iodides.

After the system had been proven to facilitate a variety of aryl iodides, the potential of applying different directing groups was investigated (Scheme 53). Other than pyridine, pyrazole as well as pyrimidine were also found qualified as directing groups in the directed carbonylative C–H bond functionalization, although pyrazole derivatives were slightly detrimental to the coupling process. Notably, heterocyclic substrates were also feasible in this catalytic system, thus 3-pyridyl-substituted furan as well as thiophene were transformed successfully to form the desired ketones in yields of 50 and 75%, respectively.

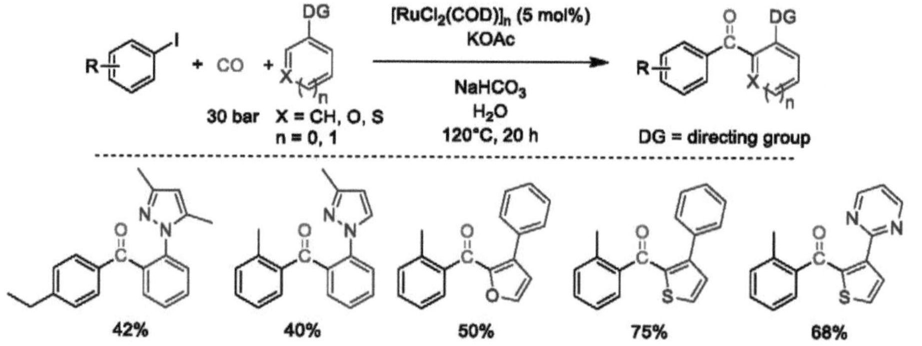

Scheme 53: Ruthenium-catalyzed carbonylative C–H functionalization with different directing groups.

Remarkably, in all experiments a high selectivity towards the monoaroylation of 2-phenyl pyridine was observed, as neither direct single nor double C–C bond formation occurred. This is in contrast to the corresponding direct coupling in the absence of CO where both mono and double functionalizations have been reported.[76]

In order to gain some details about the observed selectivity, some deuteration experiments were conducted (Scheme 54). When the carbonylative reaction was either initiated from deuterated 2-phenylpyridine or carried out in deuterated water, a significant H/D exchange was observed after 20 hours. Notably, less than 5% of H/D exchange occurred when the reaction of 2-phenyl-pyridine was stopped after five hours. Hence, it is proposed that once the product has been formed, a second metalation occurs that is reversible. Certainly, another carbonylative C–C bond formation subsequent to that second C–H bond metalation is not favored.

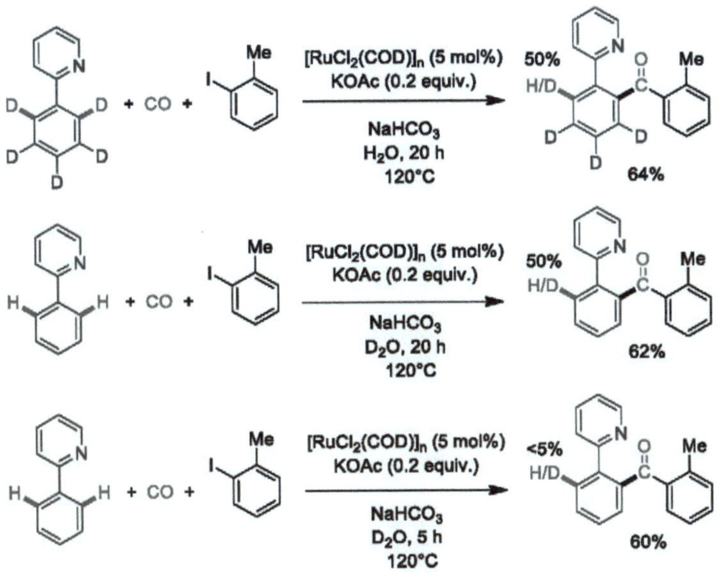

4   Summary of Results

Scheme 54: Ruthenium-catalyzed aroylation of aryl iodides – deuteration experiments.

Overall, the reported system represents the first carbonylative C–C coupling via a directed C–H bond functionalization of an arene. The application of a simple ruthenium catalyst in water as solvent enables the aroylation of 2-arylpyridines and related derivatives in good to very good yields. Superior to alternative carbonylative coupling reactions, the employment of stoichiometric amounts of organometallic reagents and the accompanied waste formation is abandoned. For further details, see publication 5.7, *Angew. Chem.* **2013**, *125*, 6413–6417; *Angew. Chem. Int. Ed.* **2013**, *52*, 6293–6297.

4.8 Hydroaroylation of Styrenes via Directed C–H Bond Activation

In order to make a first step towards a more sustainable synthesis of aromatic ketones, the aroylation of aryl iodides via a directing group assisted C–H bond activation of arenes was developed (see section 4.7).[108] However, the employment of aryl halides is still accompanied by the formation of stoichiometric amounts of salt-waste. Thus, it became of interest to make a second step towards a greener approach of producing aromatic ketones. By taking advantage of the ability of ruthenium complexes to activate C–H bonds,[73] and being inspired by ruthenium-catalyzed hydroarylations of alkenes,[109] it was aspired to perform a corresponding hydroaroylation by additionally incorporating carbon monoxide into the coupling process.[110] Since similar studies had been limited to the functionalization of ethylene or individual aliphatic alkenes as coupling partner, the carbonylative coupling of aromatic alkenes with (hetero)arenes became the central point of the following studies.

Initially, 2-phenylpyridine, styrene, and carbon monoxide were reacted in a three-component reaction in the presence of $[Ru(cod)Cl_2]_n$ polymer as the catalyst. In compliance with previous findings (see section 4.7), the selection of solvent was crucial for a decent reaction outcome. After testing common polar and non-polar organic solvents had not revealed any conversion of the starting material, water was then employed and induced a significant increase in reactivity. As a result, the desired 3-phenyl-1-(2-(pyridin-2-yl)phenyl)propan-1-one was obtained in a 73% yield. As in the previously developed system, the reaction proceeded with high selectivity towards the mono-hydroaroylation product and no traces of direct hydroarylation or double C–H bond functionalization were detected. Attempts to further

4   Summary of Results

increase the performance of the model reaction by applying additives or different ruthenium complexes did not enhance the product formation. Also, modifications of the reaction conditions such as lowering the pressure of carbon monoxide to 20 bar or decreasing the reaction temperature to 120°C proved to be detrimental on the coupling process. Hence, the simple conversion of the three coupling partners in the presence of 5 mol% of [Ru(cod)Cl$_2$]$_n$ polymer at 130°C in water emerged as optimized reaction conditions.

In order to examine the substrate scope, the carbonylation of 2-phenylpyridine with different styrenes was conducted (Scheme 55). Moderate to very good yields ranging from 38 to 78% were observed for styrenes bearing either electron-donating or -withdrawing groups, where no relation between the position of the substituent and the yield could be found. It should be noted that pentafluorostyrene was also transformed into the corresponding ketone successfully resulting in a 68% yield.

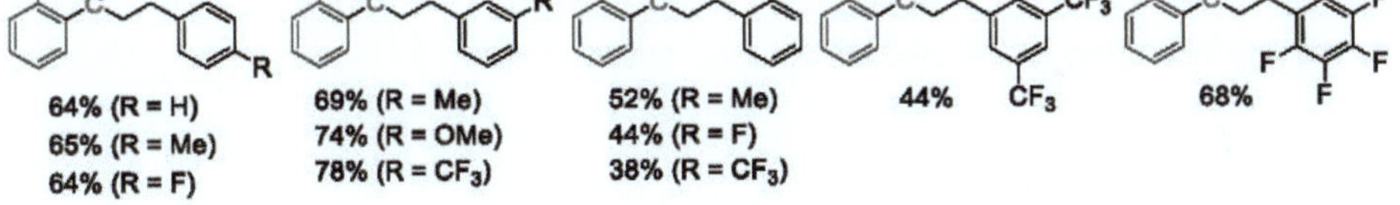

Scheme 55: Ruthenium-catalyzed hydroaroylation of different styrene derivtatives.

Other than styrenes, vinyltrimethylsilane and 1-allyl-4-fluorobenzene were also converted into the corresponding ketones in 58 and 44% yield, respectively (Scheme 56). In the latter case, two regioisomers were formed.

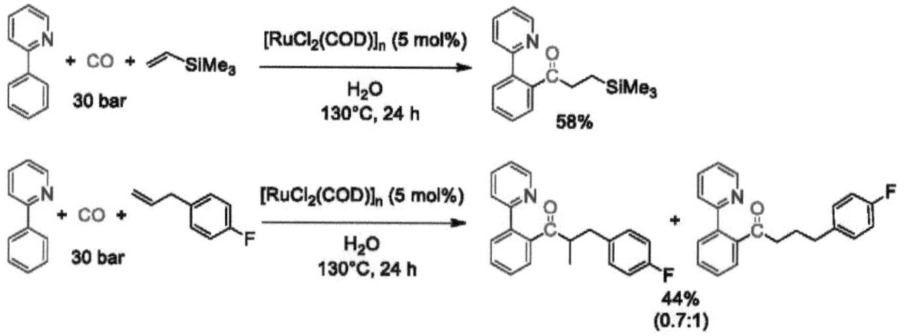

Scheme 56: Ruthenium-catalyzed hydroaroylation of alkenes.

In the following, the scope of applicable directing groups and heterocycles was explored (Scheme 57). It was found that methyl as well as methoxy substitution on the arene ring as well as on the directing group is well tolerated, generating the desired product in 66 to 94% yield. Furthermore, also pyrazole promoted the carbonylative C–H bond functionalization, although resulting in slightly lower yields of 42 and 51%. Finally, thiophene- and benzothiophene-based substrates also reacted smoothly resulting in 74 and 87% yield, respectively.

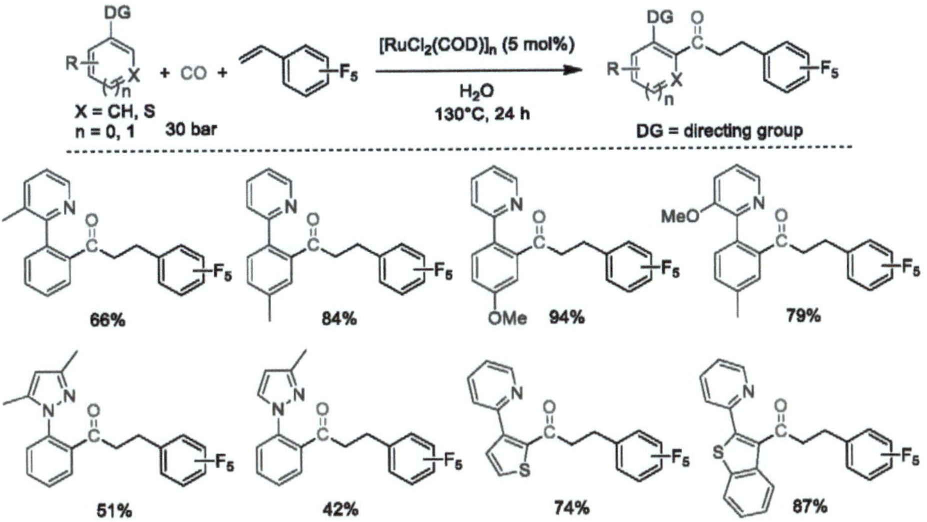

4 Summary of Results

Scheme 57: Ruthenium-catalyzed hydroaroylation of pentafluorostyrene with different directing groups.

After the vast majority of reactions had turned out to be highly selective towards monocarbonylation of one of the two *ortho*-C–H bonds of the arene, deuteration experiments were undertaken to attain further information about the reaction mechanism (Scheme 58). Employing deuterated 2-phenylpyridine in the hydroaroylation reaction led to a significant H/D exchange (80% H) in the non-functionalized *ortho*-position of the arene. Similarly H/D exchange occurred (10% D) when the reaction was carried out in deuterated water. In compliance with previous observations, a reversible metalation in the second *ortho*-position occurs, although a second carbonylation is not favored. Notably, another H/D-exchange was detected in the β-position of the carbonyl group, accounting for concurrent insertion of styrene into the Ru-D species and subsequent β-hydride elimination prior to carbon monoxide insertion and reductive elimination.

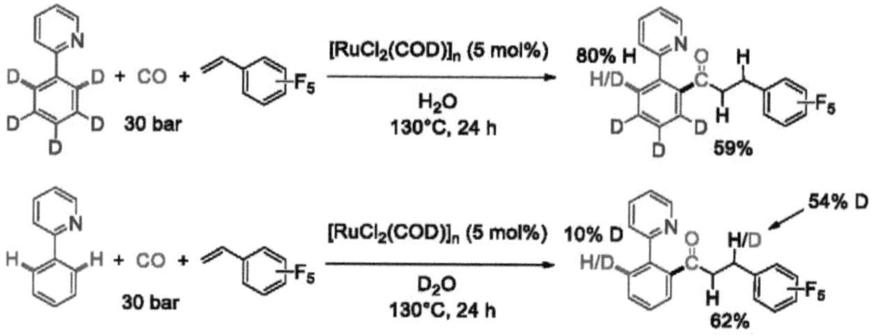

Scheme 58: Ruthenium-catalyzed hydroaroylation of pentafluorostyrene – deuteration experiments.

In conclusion, the ruthenium(II)-catalyzed carbonylative hydroarylation of styrenes via directed C–H functionalization has been reported for the first time. The three component coupling of 2-aryl(heteroaryl)pyridines, gaseous carbon monoxide and alkenes represents a 100% atom efficient process. Moreover, the reaction proceeds with high selectivity and in water as solvent, thus featuring progress towards more sustainable organic syntheses. For further details, see publication 5.8, *ChemCatChem* **2014**, *6*, 1562–1566.

4.9 Iron-Catalyzed Selective Reduction of Aldehydes under Water-Gas Shift Conditions

Besides the persistent interest in transition metal catalyzed carbonylative coupling reactions, in this thesis attention has also been attracted by further applications that make use of carbon monoxide as a cheap and abundant reagent. A famous example for such an application is the water-gas shift reaction (WGSR), since the conversion of carbon monoxide and water into carbon dioxide and hydrogen is a very important industrial process.[111] For instance, it plays a major role in the steam reforming of alkanes to produce hydrogen. In turn, it is the basis of important bulk hydrogenation processes; most notably in the Haber-Bosch process where molecular nitrogen is reduced in order to form ammonia in a Fe_3O_4-catalyzed process.[112] Obviously, the choice of the employed catalyst plays a key role in such a reaction. Traditionally, WGSRs are performed in the presence of stable heterogeneous catalysts at high temperatures and pressures, but more recently homogenous catalysts based on ruthenium, rhodium, iridium, platinum and even inexpensive iron have been found to perform the WGSR under milder reaction conditions.[111, 113] However, the direct application of carbon monoxide and water, as reductant, would represent an interesting alternative to the use of hydrogen as the reductant. Recently, the application of iron complexes as catalysts in direct hydrogenations, transfer hydrogenations and hydrosilylations, especially of carbonyl compounds, has been developed.[114]

Thus, investigations were started to conduct the reduction of unsaturated compounds with carbon monoxide and water as hydrogen source in the presence of discrete iron complexes.[115] Inspired by the work of Casey and co-workers, who used discrete cyclopentadienyliron–tricarbonyl complexes for the reduction of ketones with molecular hydrogen,[116] the cyclopentadienyliron–tricarbonyl complex was applied as a catalyst in the model reaction. Hence, benzaldehyde was reacted in the presence of 10 bar of carbon monoxide, potassium carbonate as base and the iron catalyst in water at 100°C. Instantly, reduction was observed and the corresponding benzyl alcohol was formed in a 24% yield. During the optimization process, a mixture of water and an organic solvent as the reaction medium was found to be beneficial to the reduction process. Hence, when the reaction was carried out in DMSO/H_2O (1:1), reduction occurred in 99% yield. At the same time the catalyst loading could be decreased from initially 5 to 1 mol% without any loss in conversion or yield. Next, different iron carbonyl complexes were tested in the model reaction (Scheme 59). Similarly to complex **1a**, also **1b** enabled the reduction process in a quantitative yield. Furthermore, the commercially available iron carbonyl compounds **1c**, **1d** and **1e** resulted in only traces of the reduced product, even when elevated catalyst loadings were employed. The same results were observed, when simple iron dodecacarbonyl **1f** was applied as a precatalyst. Due to its stability towards air and moisture, complex **1a** was selected for further studies.

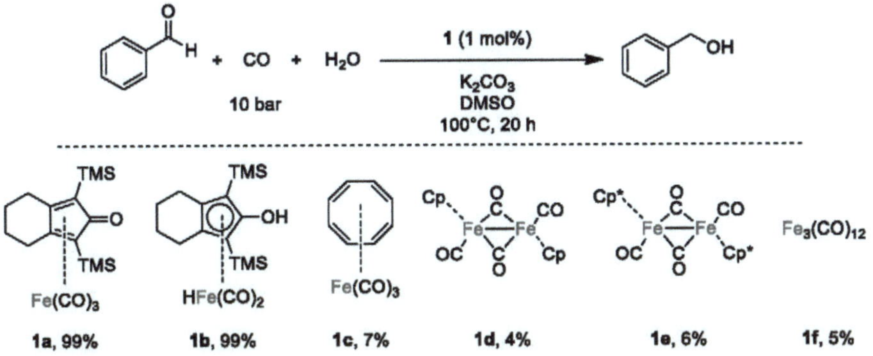

4 Summary of Results

Scheme 59: Reduction of benzaldehyde to benzyl alcohol – testing of different iron carbonyl complexes.

In order to explore the scope and limitations of the developed reduction procedure, a variety of different aromatic aldehydes were tested under the optimized reaction conditions (Scheme 60). Neither steric nor electronic properties of the aryl aldehyde showed any systematic effect on the reaction outcome since electron-donating as well as -withdrawing substituents in *ortho*-, *meta*-, and *para*-positions were well tolerated and the corresponding alcohols were formed in predominantly excellent yields. Also, functional groups such as fluoro, chloro, trifluoromethyl, methoxy as well as trifluoromethoxy substituents were not detrimental to the reduction process. However, 4-cyanobenzaldehyde resulted in a slightly diminished yield of 76%. Additionally, heterocyclic aromatic aldehydes could also be converted smoothly into the corresponding alcohols.

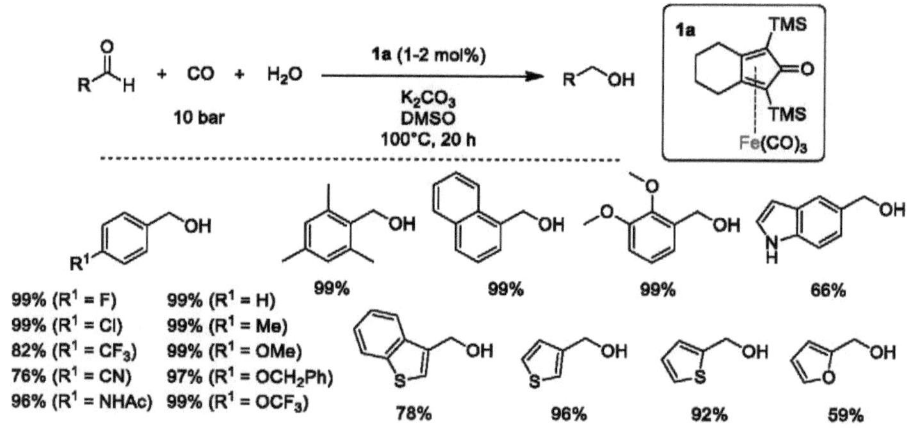

Scheme 60: Iron-catalyzed reduction of aromatic aldehydes to alcohols via WGSR.

After aromatic aldehydes had proven to be suitable substrates, the employment of more challenging aliphatic aldehydes was investigated (Scheme 61). While several cyclic and branched aliphatic aldehydes were converted into the corresponding alcohols in quantitative yields by applying only 1 mol% of the iron catalyst, the reduction of citronellal and nonanal required elevated catalyst loadings of 5 mol% to generate the corresponding alcohols in yields of 99 and 62%, respectively.

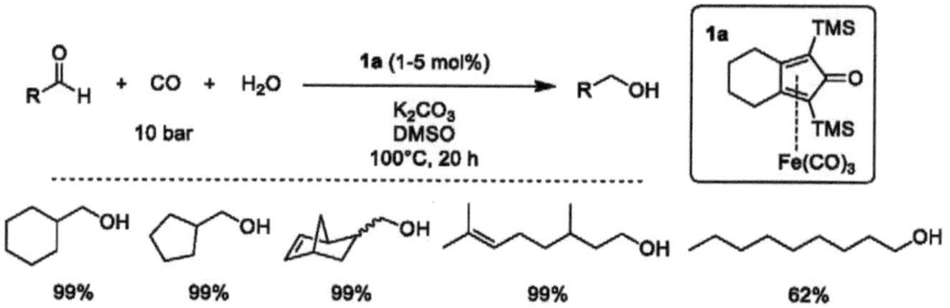

Scheme 61: Iron-catalyzed reduction of aliphatic aldehydes to alcohols via WGSR.

4 Summary of Results

Notably, when α,β-unsaturated aldehydes were reacted, both the carbonyl as well as the α,β-double bond were partly reduced. Hence, in these cases, mixtures of the resulting allylic alcohol and the fully reduced saturated alcohol were obtained.

Overall, defined iron complexes have been proven to catalyze the selective reduction of aldehydes in the presence of a base, as well as carbon monoxide and water as the reduction system. Hence, an air and moisture stable, well-defined iron complex promoted the reduction of a number of different aromatic and aliphatic aldehydes selectively and with excellent yields. Considering the good availability and low cost of both the iron catalyst and the final oxidant carbon monoxide, the developed system constitutes a valuable extension to known reports on iron catalyzed hydrogenations, transfer hydrogenations and hydrosilylations. For further details, see publication 5.9, *Chem. Eur. J.* **2012**, *18*, 15935–15939.

4.10 Conclusion

In conclusion, the toolbox of carbonylative coupling reactions has successfully been enhanced by a number of new methodologies. In doing so, a major focus of this thesis has been on the avoidance of organometallic reagents as carbon-nucleophiles.

Hence, the palladium-catalyzed carbonylative Heck reaction has been extended to the synthesis of 3-alkoxy alkenones from aryl bromides, as well as towards an unusual double carbonylative cyclization to 5-arylfuranones.

Also palladium-catalyzed is the carbonylative coupling of C–H-acidic compounds, which was also studied. Initial investigations revealed the carbonylative arylation of aryl ketones to vinylbenzoate compounds. A second approach then led to the carbonylative α-arylation of acetone and acetophenones to 1,3-diketones. In a third protocol, nitriles were employed in a carbonylative α-arylation for the first time yielding β-ketonitriles.

Furthermore, ruthenium catalyzed C–H bond activation of arenes has been found to be a suitable tool to conduct carbonylative C–C bond formations with minimum waste production. Thus, arenes bearing directing groups could successfully be reacted with aryl iodides or alkenes in the presence of carbon monoxide to generate the corresponding aromatic ketones.

In order to prove the usefulness of carbonylative coupling reactions on a real world target, bromhexine was modified in selected carbonylative Suzuki-Miura processes, as well as alkoxy and amino carbonylation reactions. The resulting compounds proved biologically active in preliminary tests.

4 Summary of Results

Finally, studies on applying carbon monoxide as a reductant were disclosed with a novel iron-catalyzed reduction of aromatic and aliphatic aldehydes to the corresponding alcohols under water-gas shift conditions.

5 Publications

5.1 Palladium-Catalyzed Carbonylative Heck Reaction of Aryl Bromides with Vinyl Ethers to 3-Alkoxy Alkenones and Pyrazoles

Johannes Schranck, Xiao-Feng Wu, Helfried Neumann, and Matthias Beller,* *Chemistry – A European Journal* **2012**, *18*, 4827–4831.

http://dx.doi.org/10.1002/chem.201103643

5.2 A Novel Double Carbonylation Reaction of Aryl Halides: Selective Synthesis of 5-Arylfuranones

Johannes Schranck, Xiao-Feng Wu, Anis Tlili, Helfried Neumann, and Matthias Beller,* *Chemistry – A European Journal* **2013**, *19*, 12959–12964.

http://dx.doi.org/10.1002/chem.201302092

5.3 A Selective Palladium-Catalyzed Carbonylative Arylation of Aryl Ketones to Give Vinylbenzoate Compounds

Johannes Schranck, Anis Tlili, Helfried Neumann, Pamela G. Alsabeh, Mark Stradiotto, and Matthias Beller,* *Chemistry – A European Journal* **2012**, *18*, 15592–15597.

http://dx.doi.org/10.1002/chem.201202895

5.4 Palladium-Catalyzed Carbonylative α-Arylation of Acetone and Acetophenones to 1,3-Diketones

Johannes Schranck, Anis Tlili, Pamela G. Alsabeh, Helfried Neumann, Mark Stradiotto, and Matthias Beller,* *Chemistry – A European Journal* **2013**, *19*, 12624–12628.

http://dx.doi.org/10.1002/chem.201302590

5.5 Palladium-Catalyzed Carbonylative α-Arylation to β-Ketonitriles

Johannes Schranck, Mia Burhardt, Christoph Bornschein, Helfried Neumann, Troels Skrydstrup,* and Matthias Beller,*
Chemistry – A European Journal **2014**, *20*, 9534–9538.

http://dx.doi.org/10.1002/chem.201402893

5.6 Palladium-Catalyzed Carbonylative Transformations of Bromhexine into Bioactive Compounds as Glucocerebrosidase Inhibitors

Muhammad Sharif, Anahit Pews-Davtyan, Jan Lukas, Johannes Schranck, Peter Langer, Arndt Rolfs,* and Matthias Beller,* *European Journal of Organic Chemistry* **2014**, 222–230.

http://dx.doi.org/10.1002/ejoc.201301180

5.7 Ruthenium-Catalyzed Carbonylative C–C Coupling in Water by Directed C–H Bond Activation

Anis Tlili, Johannes Schranck, Jola Pospech, Helfried Neumann, and Matthias Beller,* *Angewandte Chemie* **2013**, *125*, 6413–6417; *Angewandte Chemie International Edition* **2013**, *52*, 6293–6297.

http://dx.doi.org/10.1002/anie.201301663

5.8 Ruthenium-Catalyzed Hydroaroylation of Styrenes in Water via Directed C–H Bond Activation

Anis Tlili,[†] Johannes Schranck,[†] Jola Pospech, Helfried Neumann, and Matthias Beller,* *ChemCatChem* **2014**, *6*, 1562–1566.

[†]: These authors contributed equally to the work.

http://dx.doi.org/10.1002/cctc.20140203

5.9 Discrete Iron Complexes for the Selective Catalytic Reduction of Aromatic, Aliphatic, and α,β-Unsaturated Aldehydes under Water-Gas Shift Conditions

Anis Tlili, Johannes Schranck, Helfried Neumann, and Matthias Beller,*
Chemistry – A European Journal **2012**, *18*, 15935–15939.

http://dx.doi.org/10.1002/chem.201203059

6 References

[1] *Chemiewirtschaft in Zahlen 2013*, VCI - Verband der chemischen Industrie e.V., Frankfurt **2013**, pp. 32–50.

[2] a) G. Rothenberg, *Catalysis: Concepts and Green Applications*, Wiley VCH, Weinheim **2008**, pp. 1–2; b) M. Ritz, *Chemical industry contributions to coming megatrends, chemie report special 07/2012: The formula for resource efficiency*, VCI - Verband der chemischen Industrie e.V., Frankfurt **2012**, p. 7.

[3] 02/24/2014, www.chemistry2011.org; http://www.un.org/en/events/chemistry2011/.

[4] A. Behr, *Angewandte Homogene Katalyse*, Wiley VCH, Weinheim **2008**, p. 26.

[5] Dirk Steinborn, *Grundlagen der metallorganischen Komplexkatalyse*, 2nd Ed. Teubner, Wiesbaden **2010**, p. V.

[6] a) M. Beller, C. Bolm, *Transition Metals for Organic Synthesis*, Vols. 1 and 2, Wiley-VCH, Weinheim **2004**; b) J. Tsuji, *Transition Metal Reagents and Catalysts*, John Wiley & Sons, Ltd., Chichester **2000**.

[7] a) V. F. Slagt, A. H. M. de Vries, J. G. de Vries, R. M. Kellogg, *Org. Process Res. Dev.* **2010**, *14*, 30–47; b) K. C. Nicolaou, S. A. Snyder, *Classics in Total Synthesis II*, Wiley-VCH, Weinheim **2003**; c) A. B. Dounay, L. E. Overman, *Chem. Rev.* **2003**, *103*, 2945–2964; d) K. C. Nicolaou, E. J. Soerensen, *Classics in Total Synthesis*, VCH, Weinheim **1996**.

[8] For selected books on homogeneous catalysis, see: a) P. W. van Leeuwen, *Homogeneous Catalysis: Understanding the Art*, Kluwer Academic Publishers, Dordrecht **2004**; b) B. Cornils, W. A. Herrmann (Eds.), *Applied Homogeneous Catalysis with Organometallic Compounds: A Comprehensive Handbook in Three Volumes*, Wiley-VCH, Weinheim **2002**.

[9] For selected books on cross-coupling reactions, see: a) A. de Meijere, S. Bräse, M. Oestreich (Eds.), *Metal-Catalyzed Cross-Coupling Reactions and More*, Wiley-VCH, Weinheim **2014**; b) Y. Nishihara, (Ed.), *Applied Cross-Coupling Reactions*, Springer, Berlin, New York **2013**; c) N. Miyaura (Ed.), *Cross-Coupling Reactions: A Practical Guide*, Springer-Verlag, Berlin, Heidelberg **2002**.

[10] a) R. F. Heck, J. P. Nolley, *J. Org. Chem.* **1972**, *37*, 2320–2322; b) T. Mizoroki, K. Mori, A. Ozaki, *Bull. Chem. Soc. Jap.* **1971**, *44*, 581.

[11] For selected reviews, see: a) I. P. Beletskaya, A. V. Cheprakov, *Chem. Rev.* **2000**, *100*, 3009–3066; b) A. de Meijere, F. E. Meyer, *Angew. Chem.* **1994**, *106*, 2473–2506; *Angew. Chem. Int. Ed.* **1994**, *33*, 2379–2411.

[12] K. Sonogashira, Y. Tohda, N. Hagihara, *Tetrahedron Lett.* **1975**, *50*, 4467–4470.

[13] N. Miyaura, A. Suzuki, *J. Chem. Soc., Chem. Commun.* **1979**, 866–867.

[14] A. O. King, N. Okukado, E. Negishi, *J. Chem. Soc., Chem. Commun.* **1977**, 683–684.

[15] a) D. Milstein, J. K. Stille, *J. Am. Chem. Soc.* **1978**, *100*, 3636–3638; b) M. Kosugi, K. Sasazawa, Y. Shimizu, T. Migata, *Chem. Lett.* **1977**, 301–302.

[16] a) K. Tamao, K. Sumitani, M. Kumada, *J. Am. Chem. Soc.* **1972**, *94*, 4374–4376; b) R. J. P. Corriu, J. P. Massé, *J. Chem. Soc., Chem. Commun.* **1972**, 144–145.

[17] a) N. Okukado, D. E. van Horn, W. L. Klima, E. Negishi, *Tetrahedron Lett.* **1978**, *18*, 1027–1030; f) E. Negishi, N. Okukado, A. O. King, D. E. van Horn, B. I. Spiegel, *J. Am. Chem. Soc.* **1978**, *100*, 2254–2256.

[18] Y. Hatanaka, T. Hiyama, *J. Org. Chem.* **1988**, *53*, 918–920.

[19] a) T. Satoh, J. Inoh, Y. Kawamura, M. Miura, M. Nomura, *Bull. Chem. Soc. Jpn.* **1998**, *71*, 2239–2246; b) T. Satoh, Y. Kawamura, M. Miura, M. Nomura, *Angew. Chem.* **1997**, *109*, 1820–1822; *Angew. Chem. Int. Ed.* **1997**, *36*, 1740–1742.

[20] M. Palucki, S. L. Buchwald, *J. Am. Chem. Soc.* **1997**, *119*, 11108–11109.

[21] B. C. Hamann, J. F. Hartwig, *J. Am. Chem. Soc.* **1997**, *119*, 12382–12383.

[22] a) D. Solé, O. Serrano, *J. Org. Chem.* **2008**, *73*, 2476–2479; b) W. A. Moradi, S. L. Buchwald, *J. Am. Chem. Soc.* **2001**, *123*, 7996–8002.

[23] a) G. D. Vo, J. F. Hartwig, *Angew. Chem.* **2008**, *120*, 2157–2160; *Angew. Chem. Int. Ed.* **2008**, *47*, 2127–2130; b) H. Muratake, M. Natsume, H. Nakai, *Tetrahedron* **2004**, *60*, 11783–11803; c) Y. Terao, Y. Fukuoka, T. Satoh, M. Miura, M. Nomura, *Tetrahedron Lett.* **2002**, *43*, 101–104; d) H. Muratake, H. Nakai, *Tetrahedron Lett.* **1999**, *40*, 2355–2358.

[24] a) T. Hama, D. A. Culkin, J. F. Hartwig, *J. Am. Chem. Soc.* **2006**, *128*, 4976–4985; b) S. Lee, J. F. Hartwig, *J. Org. Chem.* **2001**, *66*, 3402–3415; c) K. H. Shaughnessy, B. C. Hamann, J. F. Hartwig, *J. Org. Chem.* **1998**, *63*, 6546–6553.

[25] D. A. Culkin, J. F. Hartwig, *J. Am. Chem. Soc.* **2002**, *124*, 9330–9331.

[26] For selected books on carbonylation, see: a) W. Bertleff, *Ullmann's encyclopedia of industrial chemistry*, Wiley-VCH, Weinheim, **2003**; b) L. Kollár, *Modern carbonylation methods*, Wiley-VCH, Weinheim, Chichester **2008**.

[27] For selected books and reviews, see: a) J. F. Hartwig, *Organotransition Metal Chemistry: from Bonding to Catalysis*, University Science Books, New York, **2010**; b) M. Beller, C. Bolm (Eds.) *Transition Metals for Organic Synthesis*, 2nd ed., Wiley-VCH,

6 References

Weinheim, **2004**; c) M. Beller, B. Cornils, C. D. Frohning, C. W. Kohlpaintner, *J. Mol. Catal. A* **1995**, *104*, 17–85.

[28] C. F. J. Barnard, *Organometallics* **2008**, *27*, 5402–5422.

[29] a) A. Schoenberg, I. Bartoletti, R. F. Heck, *J. Org. Chem.* **1974**, *39*, 3318–3326; b) A. Schoenberg, R. F. Heck, *J. Am. Chem. Soc.* **1974**, *96*, 7761; c) A. Schoenberg, R. F. Heck, *J. Org. Chem.* **1974**, *39*, 3327–3331.

[30] A. Brennführer, H. Neumann, M. Beller, *Angew. Chem.* **2009**, *121*, 4176–4196; *Angew. Chem. Int. Ed.* **2009**, *48*, 4114–4133.

[31] B. Cornils, W. A. Herrmann, *Applied Homogeneous Catalysis with Organometallic Compounds*, (2nd Edition), Wiley-VCH, Weinheim, **2002**, pp. 145–147.

[32] For selected examples on carbonylative Stille reactions, see: a) M. Tanaka, *Tetrahedron Lett.* **1979**, *20*, 2601–2602; b) K. Kikukawa, T. Idemoto, A. Katayama, K. Kono, F. Wada, T. Matsuda, *J. Chem. Soc. Perkin Trans. 1* **1987**, 1511–1514; c) K. Kikukawa, K. Kono, F. Wada, T. Matsuda, *Chem. Lett.* **1982**, 35–36; d) A. M. Echavarren, J. K. Stille, *J. Am. Chem. Soc.* **1988**, *110*, 1557–1565; e) H. B. Kwon, B. H. McKee, J. K. Stille, *J. Org. Chem.* **1990**, *55*, 3114–3118; f) W. J. Scott, G. T. Crisp, J. K. Stille, *Org. Synth.* **1990**, *68*, 116; g) S. D. Knight, L. E. Overman, G. Pairaudeau, *J. Am. Chem.Soc.* **1993**, *115*, 9293–9294; h) R. Shimizu, T. Fuchikami, *Tetrahedron Lett.* **1996**, *37*, 8405–8408; i) C. Chen, K. Wilcoxen, Y.-F. Zhu, K.-i. Kim, J. R. McCarthy, *J. Org. Chem.* **1999**, *64*, 3476–3482; j) S.-K. Kang, H.-C. Ryu, S.-W. Lee, *J. Organomet. Chem.* **2000**, *610*, 38–41; k) F. Garrido, S. Raeppel, A. Mann, M. Lautens, *Tetrahedron Lett.* **2001**, *42*, 265–266; l) J. Lindh, A. Fardost, M. Almeida, P. Nilsson, *Tetrahedron Lett.* **2010**, *51*, 2470–2472.

[33] For selected examples on carbonylative Sonogashira reactions, see: a) T. Kobayashi, M. Tanaka, *J. Chem. Soc., Chem. Commun.* **1981**, 333–334; b) P. G. Ciattini, E. Morera, G. Ortar, *Tetrahedron Lett.* **1991**, *32*, 6449–6452; c) L. Delaude, A. M. Masdeu, H. Alper, *Synthesis*, **1994**, 1149–1151; d) S.-K. Kang, K.-H. Lim, P.-S. Ho, W.-Y. Kim, *Synthesis*, **1997**, 874–876; e) B. Liang, M. Huang, Z. You, Z. Xiong, K. Lu, R. Fathi, J. Chen, Z. Yang, *J. Org. Chem.* **2005**, *70*, 6097–6100; f) Md. T. Rahman, T. Fukuyama, N. Kamata, M. Sato, I. Ryu, *Chem. Commun.* **2006**, 2236–2238; g) T. Fukuyama, R. Yamaura, I. Ryu, *Can. J. Chem.* **2005**, *83*, 711–715; h) V. Sans, A. M. Trzeciak, S. Luis, J. J. Ziolkowski, *Catal. Lett.* **2006**, *109*, 37–41; i) M. Iizuka, Y. Kondo, *Eur. J. Org. Chem.* **2007**, 5180–5182; j) J. Liu, J. Chen, C. Xia, *J. Catal.* **2008**, *253*, 50–56; k) X.-F. Wu, H. Neumann, M. Beller, *Chem. Eur. J.* **2010**, *16*, 12104–12107; l) X.-F. Wu, B. Sundararaju, H. Neumann, P. H. Dixneuf, M. Beller, *Chem. Eur. J.* **2011**, *17*, 106–110; m) A. Fusano, T. Fukuyama, S. Nishitani, T. Inouye, I. Ryu, *Org. Lett.* **2010**, *12*, 2410–2413.

[34] For selected examples on carbonylative Suzuki-Miura reactions, see: a) T. Ishiyama, H. Kizaki, N. Miyaura, A. Suzuki, *Tetrahedron Lett.* **1993**, *34*, 7595–7598; b) T. Ishiyama, H. Kizaki, T. Hayashi, A. Suzuki, N. Miyaura, *J. Org. Chem.* **1998**, *63*, 4726–4731; c) M. Xia, Z. Chen, *J. Chem. Res.* **1999**, 400–401; d) R. Skoda-Foeldes, Z. Szekvoelgyi, L. Kollar, Z. Berente, J. Horvath, Tuba, *Tetrahedron* **2000**, *56*, 3415–3418; e) S. Couve-Bonnaire, J.-F. Carpentier, A. Mortreux, Y. Castanet, *Tetrahedron Lett.* **2001**, *42*, 3689–3691; f) M. B. Andrus, Y. Ma, Y. Zang, C. Song, *Tetrahedron Lett.* **2002**, *43*, 9137–9140; g) D. Mingli, B. Liang, C. Wang, Z. You, J. Xiang, G. Dong, J. Chen, Z. Yang, *Adv. Synth. Catal.* **2004**, *346*, 1669–1673; h) M. Medio-Simon, C. Mollar, N. Rodriguez, G. Asensio, *Org. Lett.* **2005**, *7*, 4669–4672; i) L. Bartali, A. Guarna, P. Larini, E. G. Occhiato, *Eur. J. Org. Chem.* **2007**, 2152–2163; j) P. Prediger, A. V. Moro, C. W. Nogueira, L. Saegnago, P. H. Menezes, J. B. T. Rocha, G. Zeni, *J. Org. Chem.* **2006**, *71*, 3786–3792; k) H. Neumann, A. Brennführer, M. Beller, *Chem. Eur. J.* **2008**, *14*, 3645–3652; l) W. Qin, S. Yasuike, N. Kakusawa, J. Kurita, *J. Organomet. Chem.* **2008**, *693*, 2949–2953; m) C. Pirez, J. Dheur, M. Sauthier, Y. Castanet, A. Mortreux, *Synlett* **2009**, 1745–1748; n) P. J. Tambade, Y. P. Patil, A. G. Panda, B. M. Bhanage, *Eur. J. Org. Chem.* **2009**, 3022–3025; o) X.-F. Wu, H. Neumann, M. Beller, *Tetrahedron Lett.* **2010**, *51*, 6146–6149; p) X.-F. Wu, H. Neumann, M. Beller, *Adv. Synth. Catal.* **2011**, *353*, 788–792.

[35] For selected examples on carbonylative Negishi reactions, see: a) Y. Tamaru, H. Ochiai, Y. Yamada and Z.-i. Yoshida, *Tetrahedron Lett.* **1983**, *24*, 3869–3872; b) K. Yasui, K. Fugami, S. Tanaka, Y. Tamaru, *J. Org. Chem.* **1995**, *60*, 1365–1380; c) B. M. O'Keefe, N. Simmons, S. F. Martin, *Org. Lett.* **2008**, *10*, 5301–5304; d) Q. Wang, C. Chen, *Tetrahedron Lett.* **2008**, *49*, 2916–2921.

[36] For selected examples on carbonylative Hiyama reactions, see: a) Y. Natanaka and T. Hiyama, *Chem. Lett.* **1989**, 2049–2052; b) Y. Hatanaka, S. Fukushima, T. Hiyama, *Tetrahedron* **1992**, *48*, 2113–2126; c) S.-K. Kang, H.-C. Ryu, Y.-T. Hong, *J. Chem. Soc. Perkin Trans. 1* **2001**, 736–739.

[37] For selected examples on intramolecular carbonylative Heck reactions, see: a) E.-i. Negishi, J. A. Miller, *J. Am. Chem. Soc.* **1983**, *105*, 6761–6763; b) J. M. Tour, E.-i. Negishi, *J. Am. Chem. Soc.* **1985**, *107*, 8289–8291; c) E.-i. Negishi, G. Wu, J. M. Tour, *Tetrahedron Lett.* **1988**, *29*, 6745–7848; d) E.-i. Negishi, J. M. Tour, *Tetrahedron Lett.* **1986**, *27*, 4869–4872; e) E.-i. Negishi, S. Ma, J. Amanfu, C. Copéret, J. A. Miller, J. M. Tour, *J. Am. Chem. Soc.* **1996**, *118*, 5919–5931; f) E.-i. Negishi, C. Copéret, S. Ma, T. Mita, T. Sugihara, J. M. Tour, *J. Am. Chem. Soc.* **1996**, *118*, 5904–5918; g) C. Copéret, S. Ma, E.-i. Negishi, *Angew. Chem.* **1996**, *108*, 2255–2257; *Angew. Chem. Int. Ed. Engl.*

6 References

1996, *35*, 2125–2126; h) I. Ryu, S. Kreimerman, F. Araki, S. Nishitani, Y. Oderaotoshi, S. Minakata, M. Komatsu, *J. Am. Chem. Soc.* **2002**, *124*, 3812–3813; i) N. Chatani, A. Kamitani, M. Oshita, Y. Fukumoto, S. Murai, *J. Am. Chem. Soc.* **2001**, *123*, 12686–12687; j) S. Torii, H. Okumoto, L. H. Xu, *Tetrahedron Lett.* **1990**, *31*, 7175–7178; k) K. Okuro, M. Furuune, M. Miura, M. Nomura, *J. Org. Chem.* **1992**, *57*, 4754–4756.

[38] T. Satoh, T. Itaya, K. Okuro, M. Miura, M. Nomura, *J. Org. Chem.* **1995**, *60*, 7267–7271.

[39] a) M. A. Campo, R. C. Larock, *J. Org. Chem.* **2002**, *67*, 5616–5620; b) M. A. Campo, R. C. Larock, *Org. Lett.* **2000**, *2*, 3675–3677.

[40] C. C. C. Johansson, T. J. Colacot, *Angew. Chem.* **2010**, *122*, 686–718; *Angew. Chem. Int. Ed.* **2010**, *49*, 676–707.

[41] X.-F. Wu, P. Anbarasan, H. Neumann, M. Beller, *Angew. Chem.* **2010**, *122*, 7474–7477; *Angew. Chem. Int. Ed.* **2010**, *49*, 7316–7319.

[42] a) C. F. Barnard, *J. Org. Process Res. Dev.* **2008**, *12*, 566–574; b) C. M. Kormos, N. E. Leadbeater, *Synlett* **2007**, *13*, 2006–2010; c) R. Gaviño, S. Pellegrini, Y. Castanet, A. Mortreux, O. Mentré, *Appl. Catal. A* **2001**, *217*, 91–99.

[43] K. Okuro, H. Alper, *J. Org. Chem.* **1997**, *62*, 1566–1567.

[44] F. Ye, H. Alper, *J. Org. Chem.* **2007**, *72*, 3218–3222.

[45] K. Sangu, T. Watanabe, J. Takaya, N. Iwasawa, *Synlett* **2007**, 929–933.

[46] X.-F. Wu, H. Neumann, M. Beller, *Angew. Chem.* **2010**, *122*, 5412–5416; *Angew. Chem. Int. Ed.* **2010**, *49*, 5284–5287.

[47] a) R. J. Anto, K. Sukumaran, G. Kuttan, M. N. A. Rao, V. Subbaraju, R. Kuttan, *Cancer Lett.* **1995**, *97*, 33–37; b) B. P. Bandgar, S. S. Gawande, R. G. Bodade, J. V. Totre, C. N. Khobragade, *Bioorg. Med. Chem.* **2010**, 18, 1364–1370; c) W. M. Weber, L. A. Hunsaker, S. F. Abcouwer, L. M. Deck, D. L. V. Jagt, *Bioorg. Med. Chem.* **2005**, *13*, 3811–3820; d) W. M. Weber, L. A. Hunsaker, C. N. Roybal, E. V. Bobrovnikova-Marjon, S. F. Abcouwer, R. E. Royer, L. M. Deck, D. L. V. Jagt, *Bioorg. Med. Chem.* **2006**, *14*, 2450–2461; e) A. Modzelewska, C. Pettit, G. Achanta, N. E. Davidson, P. Huang, S. R. Khan, *Bioorg. Med. Chem.* **2006**, *14*, 3491–3495; f) Z. Nowakowska, *Eur. J. Med. Chem.* **2007**, *42*, 125–137.

[48] X.-F. Wu, H. Neumann, A. Spannenberg, T. Schulz, H. Jiao, M. Beller, *J. Am. Chem. Soc.* **2010**, *132*, 14596–14602.

[49] X.-F. Wu, H. Neumann, M. Beller, *Eur. J. Org. Chem.* **2011**, 4919–4924.

[50] X.-F. Wu, H. Jiao, H. Neumann, M. Beller, *ChemCatChem* **2011**, *3*, 726–733.

[51] P. Hermange, A. T. Lindhardt, R. H. Taaning, K. Bjerglund, D. Lupp, T. Skrydstrup, *J. Am. Chem. Soc.* **2011**, *133*, 6061–6071.

6 References

[52] P. Hermange, T. M. Gøgsig, A. T. Lindhardt, R. H. Taaning, T. Skrydstrup, *Org. Lett.* **2011**, *13*, 2444–2447.

[53] J. Schranck, X.-F. Wu, H. Neumann, M. Beller, *Chem. Eur. J.* **2012**, *18*, 4827–4831.

[54] T. M. Gøgsig, D. U. Nielsen, A. T. Lindhardt, T. Skrydstrup, *Org. Lett.* **2012**, *14*, 2536–2539.

[55] J. Schranck, X.-F. Wu, A. Tlili, H. Neumann, M. Beller, *Chem. Eur. J.* **2013**, *19*, 12959–12964.

[56] a) H. Muratake, A. Hayakawa, M. Natsume, *Tetrahedron Lett.* **1997**, *38*, 7577–7580; b) H. Muratake, M. Natsume, *Tetrahedron Lett.* **1997**, *38*, 7581–7582.

[57] For selected examples, see a) J. H. Ryan, P. J. Stang, *Tetrahedron Lett.* **1997**, *38*, 5061–5064; b) J. Morgan, J. T. Pinhey, B. A. Rowe, *J. Chem. Soc. Perkin Trans. 1* **1997**, 1005–1008; c) T. Mino, T. Matsuda, K. Maruhashi, M. Yamashita, *Organometallics* **1997**, 16, 3241–3242.

[58] For selected highlights and reviews, see: a) G. C. Lloyd-Jones, *Angew. Chem.* **2002**, *114*, 995–998; *Angew. Chem. Int. Ed.* **2002**, *41*, 953–956; b) D. A. Culkin, J. F. Hartwig, *Acc. Chem. Res.* **2003**, *36*, 234–245; c) A. C. B. Burtoloso, *Synlett* **2009**, 320–327; d) F. Bellina, R. Rossi, *Chem. Rev.* **2010**, *110*, 1082–1146.

[59] a) S. M. Crawford, P. G. Alsabeh, M. Stradiotto, *Eur. J. Org. Chem.* **2012**, 6042–6050; b) K. D. Hesp, R. J. Lundgren, M. Stradiotto, *J. Am. Chem. Soc.* **2011**, 133, 5194–5197; c) L. Ackermann, V. P. Mehta, *Chem. Eur. J.* **2012**, *18*, 10230–10233; d) P. Li, B. Lu, C. Fu, S. Ma, *Adv. Synth. Catal.* **2013**, *355*, 1255–1259; e) P. G. Alsabeh, M. Stradiotto, *Angew. Chem.* **2013**, *125*, 7383–7387; *Angew. Chem. Int. Ed.* **2013**, *52*, 7242–7246.

[60] T. Kobayashi, M. Tanaka, *Tetrahedron Lett.* **1986**, *27*, 4745–4748.

[61] E.-i. Negishi, Y. Zhang, I. Shimyama, G. Wu, *J. Am. Chem. Soc.* **1989**, *111*, 8018–8020.

[62] E.-i. Negishi, C. Coperet, T. Sugihara, I. Shimoyama, Y. Zhang, G. Wu, J. M. Tour, *Tetrahedron* **1994**, *30*, 425–436.

[63] C. Peng, J. Cheng, J. Wang, *J. Am. Chem. Soc.* **2007**, *129*, 8708–8709.

[64] T. M. Gøgsig, R. H. Taaning, A. T. Lindhardt, T. Skrydstrup, *Angew. Chem.* **2012**, *124*, 822–825; *Angew. Chem. Int. Ed.* **2012**, *51*, 798–801.

[65] S. Korsager, D. U. Nielsen, R. H. Taaning, A. T. Lindhardt, T. Skrydstrup, *Chem. Eur. J.* **2013**, *19*, 17687–17691.

[66] J. Schranck, A. Tlili, P. G. Alsabeh, H. Neumann, M. Stradiotto, M. Beller, *Chem. Eur. J.* **2013**, *19*, 12624–12628.

[67] S. Korsager, D. U. Nielsen, R. H. Taaning, T. Skrydstrup, *Angew. Chem.* **2013**, *125*, 9945–9948; *Angew. Chem. Int. Ed.* **2013**, *52*, 9763–9766.

6 References

[68] For selected recent reviews on the C–H bond functionalization of arenes, see: a) C. S. Yeung, V. M. Dong, *Chem. Rev.* **2011**, *111*, 1215–1292; b) T. C. Boorman, I. Larrosa, *Chem. Soc. Rev.* **2011**, *40*, 1910–1925; c) M. N. Hopkinson, A. D. Gee, V. Gouverneur, *Chem. Eur. J.* **2011**, *17*, 8248–8262; d) G. P. Chiusoli, M. Catellani, M. Costa, E. Motti, N. Della Ca', G. Maestri, *Coord. Chem. Rev.* **2010**, *254*, 456–469; e) G. P. McGlacken, L. M. Bateman, *Chem. Soc. Rev.* **2009**, *38*, 2447–2464; f) P. Thansandote, M. Lautens, *Chem. Eur. J.* **2009**, *15*, 5874–5883.

[69] For selected recent general reviews and highlights on C–H activation, see: a) N. Kuhl, M. N. Hopkinson, J. Wencel-Delord, F. Glorius, *Angew. Chem.* **2012**, *124*, 10382–10401; *Angew. Chem. Int. Ed.* **2012**, *51*, 10236–10254; b) J. Wencel-Delord, T. Dröge, F. Liu, F. Glorius, *Chem. Soc. Rev.* **2011**, *40*, 4740–4761; c) X. Bugaut, F. Glorius, *Angew. Chem.* **2011**, *123*, 7618–7620; *Angew. Chem. Int. Ed.* **2011**, *50*, 7479–7481.

[70] a) J. P. Djukic, J. B. Sortais, L. Barloy, M. Pfeffer, *Eur. J. Inorg. Chem.* **2009**, 817–853; b) Y. Boutadla, O. Al-Duaij, D. L. Davies, G. A. Griffith, K. Singh, *Organometallics* **2009**, *28*, 433–440; c) J. Dupont, C. S. Consorti, J. Spencer, *Chem. Rev.* **2005**, *105*, 2527–2571; d) D. L. Davies, O. Al-Duaij, J. Fawcett, M. Giardiello, S. T. Hilton, D. R. Russell, *Dalton Trans.* **2003**, 4132–4138.

[71] For selected recent reviews on palladium catalyzed C–H bond activation, see: a) E. M. Beck, M. J. Gaunt, *Top. Curr. Chem.* **2010**, *292*, 85–121; b) T. W. Lyons, M. S. Sanford, *Chem. Rev.* **2010**, *110*, 1147–1169; c) P. Sehnal, R. J. K. Taylor, I. J. S. Fairlamb, *Chem. Rev.* **2010**, *110*, 824–889; d) C. L. Sun, B. J. Li, Z. J. Shi, *Chem. Commun.* **2010**, *46*, 677–685.

[72] For selected recent reviews on rhodium catalyzed C–H bond activation, see: a) D. A. Colby, R. G. Bergman, J. A. Ellman, *Chem. Rev.* **2010**, *110*, 624–655; b) J. Bouffard, K. Itami, *Top. Curr. Chem.* **2010**, *292*, 231–280; c) J. C. Lewis, R. G. Bergman, J. A. Ellman, *Acc. Chem. Res.* **2008**, *41*, 1013–1025; d) K. Fagnou, M. Lautens, *Chem. Rev.* **2003**, *103*, 169–196; e) H. M. L. Davies, R. E. J. Beckwith, *Chem. Rev.* **2003**, *103*, 2861–2903.

[73] For selected recent reviews on ruthenium catalyzed C–H bond activation, see: a) L. Ackermann, *Chem. Rev.* **2011**, *111*, 1315–1345; b) L. Ackermann, R. Vicente, *Top. Curr. Chem.* **2010**, *292*, 211–229; c) L. Ackermann, *Chem. Commun.* **2010**, *46*, 4866–4877; d) L. Ackermann, R. Vicente, A. R. Kapdi, *Angew. Chem.* **2009**, *121*, 9976–10011; *Angew. Chem. Int. Ed.* **2009**, *48*, 9792–9826.

[74] a) Q. Liu, H. Zhang, A. Lei, *Angew. Chem.* **2011**, *123*, 10978–10989; *Angew. Chem. Int. Ed.* **2011**, *50*, 10788–10799; b) X.-F. Wu, H. Neumann, *ChemCatChem* **2012**, *4*, 447–458.

[75] H. Zhang, R. Shi, P. Gan, C. Liu, A. Ding, Q. Wang, A. Lei, *Angew. Chem.* **2012**, *124*, 5294–5297; *Angew. Chem. Int. Ed.* **2012**, *51*, 5204–5207.

[76] P. B. Arockiam, C. Bruneau, P. H. Dixneuf, *Chem. Rev.* **2012**, *112*, 5879–5918.

[77] For selected examples, see: a) N. Chatani, S. Inoue, K. Yokota, H. Tatamidani, Y. Fukumoto, *Pure Appl. Chem.* **2010**, *82*, 1443–1451; b) S. Inoue, H. Shiota, Y. Fukumoto, N. Chatani, *J. Am. Chem. Soc.* **2009**, *131*, 6898–6899.

[78] E. J. Moore, W. R. Pretzer, T. J. OConnell, J. Harris, L. LaBounty, L. Chou, S. S. Grimmer, *J. Am. Chem. Soc.* **1992**, *114*, 5888–5890.

[79] N. Chatani, Y. Ie, F. Kakiuchi, S. Murai, *J. Org. Chem.* **1997**, *62*, 2604–2610.

[80] S. Imoto, T. Uemura, F. Kakiuchi, N. Chatani, *Synlett* **2007**, 170–172.

[81] For recent mechanistic investigations on Ru-catalyzed C–H activation processes, see: a) I. Fabre, N. von Wolff, G. Le Duc, E. F. Flegeau, C. Bruneau, P. H. Dixneuf, A. Jutand, *Chem. Eur. J.* **2013**, *19*, 7595–7604; b) E. F. Flegeau, C. Bruneau, P. H. Dixneuf, A. Jutand, *J. Am. Chem. Soc.* **2011**, *133*, 10161–10170; c) L. Ackermann, R. Vicente, H. K. Potukuchi, V. Pirovano, *Org. Lett.* **2010**, *12*, 5032–5035.

[82] T. Fukuyama, N. Chatani, J. Tatsumi, F. Kakiuchi, S. Murai, *J. Am. Chem. Soc.* **1998**, *120*, 11522–11523.

[83] a) Y. Ie, N. Chatani, T. Ogo, D. R. Marshall, T. Fukuyama, F. Kakiuchi, S. Murai, *J. Org. Chem.* **2000**, *65*, 1475–1488; b) T. Asaumi, N. Chatani, T. Matsuo, F. Kakiuchi, S. Murai, *J. Org. Chem.* **2003**, *68*, 7538–7540.

[84] N. Chatani, S. Yorimistu, T. Asaumi, F. Kakiuchi, S. Murai, *J. Org. Chem.* **2002**, *67*, 7557–7560.

[85] For recent examples on using 3-alkoxy-alkenones as building blocks, see: a) G. Mross, S. Ladzik, H. Reinke, P. Langer, A. Spannenberg, C. Fischer, *Synthesis* **2009**, 2236–2248; b) G. Mross, H. Reinke, P. Langer, *Synlett* **2008**, 963–966; c) E. Stern, G. G. Muccioli, B. Bosier, L. Hamtiaux, R. Millet, J. H. Poupaert, J.-P. Henichart, P. Depreux, J.-F. Goossens, D. M. Lambert, *J. Med. Chem.* **2007**, *50*, 5471–5484.

[86] For selected recent examples, see: a) S. Fustero, R. Román, J. F. Sanz-Cervera, A. Simón-Fuentes, J. Bueno, S. Villanova, *J. Org. Chem.* **2008**, *73*, 8545–8552; b) T. de Paulis, K. Hemstapat, Y. Chen, Y. Zhang, S. Saleh, D. Alagille, R. M. Baldwin, G. D. Tamagnan, P. J. Conn, *J. Med. Chem.* **2006**, *49*, 3332–3344; c) Y. Shiga, I. Okada, Y. Ikeda, E. Takizawa, T. J. Fukuchi, *Pesticide Sci.* **2003**, *28*, 313–314.

[87] a) E. E. Shults, J. Velder, H.-G. Schmalz, S. V. Chernov, T. V. Rubalava, Y. V. Gatilov, G. Henze, G. A. Tolstikov, A. Prokop, *Bioorg. Med. Chem. Lett.* **2006**, *16*, 4228–4231; b)

6 References

K. A. Koo, M. K. Lee, S. H. Kim, E. J. Jeong, T. H. Oh, Y. C. Kim, *Br. J. Pharmacol.* **2009**, *150*, 65–71.

[88] M. P. Kaushik, D. Thavaselvam, M. Nivsarkar, B. N. Acharya, S. Prasanna, K. Sekhar, PCT Int. Appl. (**2007**) WO 2007039915 (CA: 2007, 146, 408201).

[89] J. K. Son, D. H. Kim, M. H. Woo, *J. Nat. Prod.* **2003**, *66*, 1369–1372.

[90] H. Kikuchi, Y. Tsukitani, H. Nakanishi, I. Shimizu, S. Saitoh, K. Iguchi, Y. Yamada, *Chem. Pharm. Bull.* **1983**, *31*, 1172–1176.

[91] a) I. Sapountzis, W. Dohle, P. Knochel, *Chem. Commun.* **2001**, 2068–2069; b) P. Brownbridge, E. Egert, P. G. Hunt, O. Kennard, S. Warren, *J. Chem. Soc. Perkin Trans. 1* **1981**, 2751–2759; c) E. J. Corey, G. Schmidt, *Tetrahedron Lett.* **1980**, *21*, 731–734.

[92] For selected recent examples, see: a) L. J. Gooßen, J. Paetzold, *Angew. Chem.* **2004**, *116*, 1115–1118; *Angew. Chem. Int. Ed.* **2004**, *43*, 1095–1098; b) M. T. Reetz, L. J. Gooßen, A. Meiswinkel, J. Paetzold, J. Feldthusen Jensen, *Org. Lett.* **2003**, *5*, 3099–3101; c) P. H. Dixneuf, C. Bruneau, S. Dérien, *Pure Appl. Chem.* **1998**, *70*, 1065–1070.

[93] H. Suematsu, S. Kanchiku, T. Uchida, T. Katsuki, *J. Am. Chem. Soc.* **2008**, *31*, 10327–10337.

[94] a) A. V. Kel'in, *Curr. Org. Chem.* **2003**, *7*, 1691–1711; b) A. V. Kel'in, A. Maioli, *Curr. Org. Chem.* **2003**, *7*, 1855–1886.

[95] J. Schranck, A. Tlili, H. Neumann, P. G. Alsabeh, M. Stradiotto, M. Beller, *Chem. Eur. J.* **2012**, *18*, 15592–15597.

[96] a) F. Gosselin, P. D. O'Shea, R. A. Webster, R. A. Reamer, R. D. Tillyer, E. J. J. Grabowski, *Synlett* **2006**, 3267–3270; b) M. Bagley, M. C. Lubinu, C. Mason, *Synlett* **2007**, 0704–0708; c) Y. Luo, P. G. Potvin, *J. Org. Chem.* **1994**, *59*, 1761–1765; d) A. N. Kost, I. Grandberg, *Advances in Heterocyclic Chemistry*, Vol. 6 (Eds.: A. R. Katrizky, A. J. Boulton), Academic Press, New York, **1966**, pp. 347–429.

[97] For selected recent examples, see: a) R. W. Nowill, T. J. Patel, D. L. Beasley, J. A. Alvarez, E. Jackson, T. J. Hizer, I. Ghiviriga, S. C. Mateer, B. D. Feske, *Tetrahedron Lett.* **2011**, *52*, 2440–2441; b) O. Soltani, M. A. Arger, H. Vázquez-Villa, E. M. Carreira, *Org. Lett.* **2010**, *12*, 2893–2895; c) D. Zhu, H. Ankati, C. Mukherjee, Y. Yang, E. R. Biehl, L. Hua, *Org. Lett.* **2007**, *9*, 2561–2563; d) H. Ankatik, D. Zhu, Y. Yang, E. R. Biehl, L. Hua, *J. Org. Chem.* **2009**, *74*, 1658–1662.

[98] a) A. Park, S. Lee, *Org. Lett.* **2012**, *14*, 1118–1121; b) D. Rao, F. A. Stuber, *Synthesis* **1983**, 308.

[99] a) J. You, J. G. Verkade, *Angew. Chem.* **2003**, *115*, 5205–5207; *Angew. Chem. Int. Ed.* **2003**, *42*, 5051–5053; b) J. You, J. G. Verkade, *J. Org. Chem.* **2003**, *68*, 8003–8007; c) L. Wu, J. F. Hartwig, *J. Am. Chem. Soc.* **2005**, *127*, 15824–15832.

[100] C. Bornschein, S. Werkmeister, K. Junge, M. Beller, *New J. Chem.* **2013**, *37*, 2061–2065.

[101] a) R. Giridhar, A. K. Mishra, *Bioorg. Med. Chem.* **2008**, *16*, 9443–9449; b) I. Andreu, I. M. Morera, F. Bosca, L. Sanchez, P. Camps, M. A. Miranda, *Org. Biomol. Chem.* **2008**, *6*, 860–867; c) J. Norinder, K. Bogar, L. Kanupp, J.-E. Baeckvall, *Org. Lett.* **2007**, *9*, 5095–5098; d) A. Herschhorn, L. Lerman, M. Weitman, I. O. Gleenberg, A. Nudelman, A. Hizi, *J. Med. Chem.* **2007**, 50, 2370–2384.

[102] M. Sharif, A. Pews-Davtyan, J. Lukas, J. Schranck, P. Langer, A. Rolfs, M. Beller, *Eur. J. Org. Chem.* **2014**, 222–230.

[103] B. Latli, M. Hrapchak, H.-K. Switeck, D. M. Retz, D. Krishnamurthy, C. H. Senanayake, *J. Labelled Compd. Radiopharm.* **2010**, *53*, 15–23.

[104] For selected recent examples, see: a) J.-F. Gremmel, *J. Vet. Pharmacol. Ther.* **2010**, *27*, 219–225; b) K. Nobata, M. Fujimura, Y. Ishiura, S. Myou, S. Nakao, *Clin. Exp. Med.* **2006**, *6*, 79–83; c) L. A. Meijer, J. C. M. Verstegen, S. Bull, J. Fink-Gremmels, *J. Vet. Pharmacol. Ther.* **2004**, *27*, 219–225.

[105] a) I. Bendikov-Bar, I. Ron, M. Filocamo, M. Horowitz, *Blood Cells Mol. Dis.* **2011**, *46*, 4–10; b) G. H. Maegawa, M. B. Tropak, J. D. Buttner, B. A. Rigat, M. Fuller, D. Pandit, L. Tang, G. J. Kornhaber, Y. Hamuro, J. T. Clarke, D. J. Mahuran, *J. Biol. Chem.* **2009**, *284*, 23502–23516.

[106] N. A. McGrath, M. Brichacek, J. T. Njardarson, *J. Chem. Educ.*, **2010**, *87*, 1348–1349.

[107] a) P. B. Arockiam, C. Fischmeister, C. Bruneau, P. H. Dixneuf, *Green Chem.* **2013**, *15*, 67–71; b) P. B. Arockiam, C. Fischmeister, C. Bruneau, P. H. Dixneuf, *Angew. Chem.* **2010**, *122*, 6779–6782; *Angew. Chem. Int. Ed.* **2010**, 49, 6629–6632; c) F. Požgan, P. H. Dixneuf, *Adv. Synth. Catal.* **2009**, *351*, 1737–1743.

[108] A. Tlili, J. Schranck, J. Pospech, H. Neumann, M. Beller, *Angew. Chem.* **2013**, *125*, 6413–6417; *Angew. Chem. Int. Ed.* **2013**, *52*, 6293–6297.

[109] a) F. Kakiuchi, T. Tsujimoto, M. Sonoda, N. Chatani, S. Murai, *Synlett* **2001**, 948–951; b) F. Kakiuchi, M. Sonoda, T. Tsujimoto, N. Chatani, S. Murai, *Chem. Lett.* **1999**, 1083–1084; c) F. Kakiuchi, T. Sato, M. Yamauchi, N. Chatani, S. Murai, *Chem. Lett.* **1999**, 19–20; d) F. Kakiuchi, T. Sato, T. Tsujimoto, M. Yamauchi, N. Chatani, S. Murai, *Chem. Lett.* **1998**, *27*, 1053–1054; e) F. Kakiuchi, M. Yamauchi, N. Chatani, S. Murai, *Chem. Lett.* **1996**, 111–112.

[110] A. Tlili, J. Schranck, J. Pospech, H. Neumann, M. Beller, *ChemCatChem* **2014**, *6*, 1562–1566.

6 References

[111] For selected reviews, see: a) A. J. Esswein, D. G. Nocera, *Chem. Rev.* **2007**, *107*, 4022–4047; b) R. B. King, *J. Organomet. Chem.* **1999**, 586, 2–17; c) P. C. Ford, *Acc. Chem. Res.* **1981**, *14*, 31–37.

[112] a) J. R. Jennings (Ed.), *Catalytic Ammonia Synthesis: Fundamentals and Practice*, Plenum, New York, **1991**; b) R. Schlögl, *Angew. Chem.* **2003**, *115*, 2050–2055; *Angew. Chem. Int. Ed.* **2003**, *42*, 2004–2008; c) M. Appel, *Ullmann's Encyclopedia of Industrial Chemistry: Ammonia*, Wiley-VCH, Weinheim, **2011**.

[113] R. B. King, C. C. Frazier, R. M. Hanes, A. D. King Jr., *J. Am. Chem. Soc.* **1978**, *100*, 2925–2927.

[114] For selected recent reviews and highlights, see: a) H. Nakazawa, M. Itazaki, *Top. Organomet. Chem.* **2011**, *33*, 27–81; b) C.-L. Sun, B.-J. Li, Z.- J. Shi, *Chem. Rev.* **2011**, *111*, 1293–1314; c) K. Junge, K. Schröder, M. Beller, *Chem. Commun.* **2011**, *47*, 4849–4859; d) W. M. Czaplik, M. Mayer, J. Cvengros, A. Jacobi von Wangelin, *ChemSusChem* **2009**, *2*, 396–417; e) R. H. Morris, *Chem. Soc. Rev.* **2009**, *38*, 2282–2291; f) E. B. Bauer, *Curr. Org. Chem.* **2008**, *12*, 1341–1369; g) S. Enthaler, K. Junge, M. Beller, *Angew. Chem.* **2008**, *120*, 3363–3367; *Angew. Chem. Int. Ed.* **2008**, *47*, 3317–3321; h) A. Correa, O. G. Mancheño, C. Bolm, *Chem. Soc. Rev.* **2008**, *37*, 1108–1117.

[115] A. Tlili, J. Schranck, H. Neumann, M. Beller, *Chem Eur. J.* **2012**, *18*, 15935–15939.

[116] C. P. Casey, H. Guan, *J. Am. Chem. Soc.* **2007**, *129*, 5816–5817.

yes
I want morebooks!

Buy your books fast and straightforward online - at one of the world's fastest growing online book stores! Environmentally sound due to Print-on-Demand technologies.

Buy your books online at
www.get-morebooks.com

Kaufen Sie Ihre Bücher schnell und unkompliziert online – auf einer der am schnellsten wachsenden Buchhandelsplattformen weltweit! Dank Print-On-Demand umwelt- und ressourcenschonend produziert.

Bücher schneller online kaufen
www.morebooks.de

OmniScriptum Marketing DEU GmbH
Heinrich-Böcking-Str. 6-8
D - 66121 Saarbrücken

Telefax: +49 681 93 81 567-9

info@omniscriptum.de
www.omniscriptum.de

Printed by Books on Demand GmbH, Norderstedt / Germany